# Communications
# in Computer and Information Science 2328

Series Editors

Gang Li , *School of Information Technology, Deakin University, Burwood, VIC, Australia*
Joaquim Filipe , *Polytechnic Institute of Setúbal, Setúbal, Portugal*
Zhiwei Xu, *Chinese Academy of Sciences, Beijing, China*

## Rationale

The CCIS series is devoted to the publication of proceedings of computer science conferences. Its aim is to efficiently disseminate original research results in informatics in printed and electronic form. While the focus is on publication of peer-reviewed full papers presenting mature work, inclusion of reviewed short papers reporting on work in progress is welcome, too. Besides globally relevant meetings with internationally representative program committees guaranteeing a strict peer-reviewing and paper selection process, conferences run by societies or of high regional or national relevance are also considered for publication.

## Topics

The topical scope of CCIS spans the entire spectrum of informatics ranging from foundational topics in the theory of computing to information and communications science and technology and a broad variety of interdisciplinary application fields.

## Information for Volume Editors and Authors

Publication in CCIS is free of charge. No royalties are paid, however, we offer registered conference participants temporary free access to the online version of the conference proceedings on SpringerLink (http://link.springer.com) by means of an http referrer from the conference website and/or a number of complimentary printed copies, as specified in the official acceptance email of the event.

CCIS proceedings can be published in time for distribution at conferences or as post-proceedings, and delivered in the form of printed books and/or electronically as USBs and/or e-content licenses for accessing proceedings at SpringerLink. Furthermore, CCIS proceedings are included in the CCIS electronic book series hosted in the SpringerLink digital library at http://link.springer.com/bookseries/7899. Conferences publishing in CCIS are allowed to use Online Conference Service (OCS) for managing the whole proceedings lifecycle (from submission and reviewing to preparing for publication) free of charge.

## Publication process

The language of publication is exclusively English. Authors publishing in CCIS have to sign the Springer CCIS copyright transfer form, however, they are free to use their material published in CCIS for substantially changed, more elaborate subsequent publications elsewhere. For the preparation of the camera-ready papers/files, authors have to strictly adhere to the Springer CCIS Authors' Instructions and are strongly encouraged to use the CCIS LaTeX style files or templates.

## Abstracting/Indexing

CCIS is abstracted/indexed in DBLP, Google Scholar, EI-Compendex, Mathematical Reviews, SCImago, Scopus. CCIS volumes are also submitted for the inclusion in ISI Proceedings.

## How to start

To start the evaluation of your proposal for inclusion in the CCIS series, please send an e-mail to ccis@springer.com.

Mirko Presser · Antonio Skarmeta · Srdjan Krco ·
Aurora González Vidal
Editors

# Global Internet of Things and Edge Computing Summit

First Global Summit, GIECS 2024
Brussels, Belgium, September 24–25, 2024
Proceedings

 Springer

*Editors*
Mirko Presser ⓘ
Centre for Business Development
Aarhus University
Herning, Denmark

Srdjan Krco ⓘ
DunavNET
Novi Sad, Serbia

Antonio Skarmeta ⓘ
Department of Information
and Communication Engineering
University of Murcia
Murcia, Spain

Aurora González Vidal ⓘ
Department of Information
and Communication Engineering
University of Murcia
Murcia, Spain

ISSN 1865-0929          ISSN 1865-0937 (electronic)
Communications in Computer and Information Science
ISBN 978-3-031-78571-9          ISBN 978-3-031-78572-6 (eBook)
https://doi.org/10.1007/978-3-031-78572-6

This work was supported by Alliance for IoT and Edge Computing Innovation IVZW.

This Springer imprint is published by the registered company Springer Nature Switzerland AG
The registered company address is: Gewerbestrasse 11, 6330 Cham, Switzerland

If disposing of this product, please recycle the paper.

# Preface

This volume comprises the select proceedings of the Global IoT and Edge Computing Summit (GIECS) 2024. Its content will be of interest to academics and industry professionals working in computation, communication, and engineering.

GIECS is an international conference established to attract and present the latest research findings in IoT and Edge Computing. Through a systematic peer-review process, contributions from researchers, developers, and practitioners working at the intersection of cloud, edge and IoT were evaluated, and the top papers selected, to foster a deeper understanding of how continuum computing is shaping the future of Internet.

The conference invited submissions on a range of topics, including but not limited to:

- IoT Enabling Technologies
- IoT Applications, Services and Real Implementations
- End-User and Human-Centric IoT, including IoT Multimedia, Societal Impacts and Sustainable Development
- IoT Security, Privacy and Data Protection
- IoT Pilots, Testbeds and Experimentation Results

The 12 full papers presented in these proceedings were carefully reviewed and selected from 21 submissions, each of which underwent a single-blind peer-review process. Each paper was reviewed in two rounds, with feedback provided by three independent domain experts. This thorough review ensured that only the highest-quality research was included in the final selection. To maintain the highest standards of academic integrity and impartiality, we ensured that contributions co-authored by members of the GIECS 2024 committees were reviewed with strict adherence to a transparent and objective process.

Of the 12 accepted papers, three were co-authored by committee members. To prevent any conflict of interest, the following measures were implemented:

- Blind Review Process: All submissions, including those co-authored by committee members, underwent a single-blind review, where authors were unaware of the reviewers' identities. This ensured that papers were evaluated solely on their scholarly merit.
- Independent Reviewers: Committee members who were co-authors of any submission were entirely excluded from both the selection of reviewers and the decision-making process for their papers. Independent domain experts, unaffiliated with the authors, were selected to review these submissions.

By implementing these measures, we ensured that all accepted papers, regardless of authorship, were judged fairly and in accordance with the conference's review criteria. Hence, the papers featured in this volume were selected based on the quality and impact of their research, without any bias or preferential treatment.

We extend our deepest gratitude to all the contributors, whose dedication and insightful work made GIECS 2024 a success. The event served as a collaborative platform for exchanging ideas, sharing research findings and collaborating on innovative solutions to advance the field of cloud-IoT-edge.

The conference took place in Brussels, Belgium, in collaboration with the Alliance for IoT and Edge Computing Innovation (AIOTI) as part of the annual event called AIOTI Days 2024, 24-25 September 2024. The event featured keynote speeches, scientific presentations, thematic workshops and panel discussions, bringing together over 120 researchers from diverse fields to address challenges in emerging disciplines. We look forward to future editions of the GIECS conference.

September 2024

Mirko Presser
Antonio Skarmeta
Srdjan Krco
Aurora González Vidal

# Organization

## General Chair

Antonio Skarmeta — University of Murcia, Spain

## Co-chairs

Mirko Presser — Aarhus University, Denmark
Srdjan Krco — DunavNET, Serbia

## Program Committee Chairs

Mirko Presser — Aarhus University, Denmark
Aurora Gonzalez Vidal — University of Murcia, Spain

## Additional Reviewers

Alberto Robles Enciso
Christian Kloch
David Sarabia
Francois Fischer
Harris Niavis
Jose Manuel Bernabe
Karagiannis Vasileios
Konstantinos Loupos
Maria Hernandez Padilla
Parwinder Singh
Valentina Tomat
Marcin Kotlinski
María Hernández Padilla
Marianne Marot
Marta Pinzone
Michail J. Beliatis

Natalia Borgonos Garcia
Nima Rahmani Choubeh
Nirvana Meratnia
Parwinder Singh
Qize Guo
Raúl García-Castro
Sergio Gusmeroli
Sokratis Vavilis
Syed Danish Abbas
Tarik Taleb
Thanasis G. Papaioannou
Walter Quadrini
Xavi Masip-Bruin
Yan Chen
Yvan Martzluff

# Contents

## Industrial Internet of Things (IIoT) and Digital Twins

## Data Management, Privacy, and Trust in Distributed Systems

## Edge Computing and Cross-Domain Systems

x      Contents

# Industrial Internet of Things (IIoT) and Digital Twins

# Semantic-Enhanced Digital Twin
# for Industrial Working Environments

Chao Yang[1](✉) ⓘ, Qize Guo[2] ⓘ, Hao Yu[2] ⓘ, Yan Chen[2] ⓘ, Tarik Taleb[3] ⓘ,
and Kari Tammi[1] ⓘ

[1] Mechanical Engineering, Aalto University, 02150 Espoo, Finland
chao.1.yang@aalto.fi
[2] ICTficial Oy, 02130 Espoo, Finland
[3] Electrical Engineering and Information Technology, Ruhr University Bochum,
44801 Bochum, Germany

**Abstract.** Real-time data from diverse Internet of Things (IoT) sensors
(such as cameras, temperature, light, and air quality sensors) is essential for monitoring smart manufacturing environments. However, efficiently perceiving, integrating, and interpreting this data remains a challenge, as it involves dealing with heterogeneous data formats, ensuring
data accuracy, and providing real-time analytics. This paper proposes a
semantic-enhanced Digital Twin (DT) to address these complexities and
aims to offer a comprehensive view of industrial working environments.
The paper first presents a conceptual overview of the semantic-enhanced
DT architecture, followed by a detailed description of the system architecture, encompassing edge, cloud, and interface modules. Additionally,
the implementation of the entire system is presented. The results demonstrate the feasibility of the proposed DT, showing its potential for deployment in real-world scenarios.

**Keywords:** Digital Twin · Industrial working environment · Internet
of Things · Semantic model

## 1 Introduction

Industry 4.0 is driving a significant digital transformation in traditional manufacturing, ushering in a new era of intelligent, flexible, and adaptable production
paradigms [1]. This shift necessitates efficient and comprehensive environmental
data perception [2]. Accurate real-time environmental data is crucial for monitoring and optimizing industrial processes. Manufacturers can adapt swiftly
to changes by gaining real-time insights into environmental conditions, ensuring optimal performance while minimizing environmental impact. This leads to
improved resource utilization, reduced waste, and increased overall efficiency.

While the Internet of Things (IoT) has revolutionized data collection in factories, enabling real-time monitoring and control for optimized production [3],
a critical gap remains. Current limitations include the lack of an efficient approach for capturing and translating real-world factory environmental data into

© The Author(s) 2025
M. Presser et al. (Eds.): GIECS 2024, CCIS 2328, pp. 3–20, 2025.
https://doi.org/10.1007/978-3-031-78572-6_1

a unified digital format. This challenge is further complicated by the inherent heterogeneity of environmental data at the factory floor level [4]. Sensor readings and environmental conditions are often measured in diverse formats and units, hindering seamless integration and analysis within a digital system. Additionally, inconsistent data quality and latency issues can result in inaccurate models and suboptimal decision-making [5]. Fragmented data silos across different devices and systems also impede the holistic view required for comprehensive analysis. Therefore, smart factories need an efficient approach to data perception and transmission, as well as a standardized method to represent and interpret this rich environmental data.

Digital Twins (DTs) can be a promising solution to address these issues by providing a cohesive, real-time digital representation of the physical environment. This allows for continuous monitoring and analysis of the factory floor, with real-time updates of environmental data. However, existing DTs often struggle with semantic interoperability [6], limiting their ability to fully leverage the rich data available. Diverse data sources and varying terminologies can lead to misunderstandings and misalignments in data interpretation. Semantic technologies establish a uniform interpretation of data, ensuring universal comprehension regardless of its source [7]. Integrating these technologies with DTs enables seamless communication and data exchange across different systems and platforms [8]. Semantic-enhanced DTs can achieve a higher level of interoperability, ensuring that data from various sources is accurately interpreted and applied.

This work proposes a novel semantic-enhanced Digital Twin (DT) for a holistic digital representation of industrial working environments. To facilitate knowledge representation and reasoning about edge devices and their perception, an ontological model is introduced. The proposed DT architecture consists of three key components: the edge module, the central module, and the interface module. The edge module is responsible for real-time environmental data perception and transmission. The central module utilizes a microservices-based architecture to enable scalable service deployment and distributed computing resources. The interface module provides a comprehensive suite of functionalities for end users. Additionally, this paper details the hardware design of the edge device and the microservices-based architecture within the central server. Finally, the system's performance is evaluated through practical validation, focusing on information retrieval capabilities, User Interface (UI) usability, and latency performance.

The paper is structured as follows: Sect. 2 provides an overview of relevant research on IoT, DTs, and semantic technologies. Section 3 introduces an ontological model and details the system architecture of the proposed DT. Section 4 covers the hardware design of the edge device, the information model it utilizes, and the microservices-based architecture employed by the central server platform. Section 5 presents the evaluation and results. Section 6 analyzes the research contributions and limitations of the proposed system. Finally, the paper concludes with a summary of the study and outlines potential avenues for future research.

## 2   Related Works

### 2.1   IoT-Based Industrial Systems

The IoT with prevalent sensing capabilities in manufacturing has transformed physical entities and operators into 'cyber-ones' [9]. By leveraging IoT technologies, industrial systems enable the monitoring and optimization of various aspects of industrial processes. Through the integration of sensors, data acquisition modules, and analytical algorithms, these systems continuously collect and analyze data from critical points within industrial operations. For instance, Hossain et al. [10] presented a novel algorithm embedded in the proposed PI-based controller for real-time fault detection in power converters. In addition to monitoring machinery and equipment for operational efficiency and downtime reduction, scholars have also focused on ensuring sustainability and safety in industrial operations. For example, Palazon et al. [11] employed Wireless Sensor Networks (WSN) with mobile motes carried by workers and vehicles within the smart factory to enhance mutual perception, thereby improving health and safety conditions in the industrial environment. Moreover, Ahn et al. [12] presented an intelligent camera-based system for managing the safety of factory environments. The integration of multiple IoT devices has become indispensable for comprehensive perception and understanding within the factory field. However, the diverse data formats and heterogeneous nature of IoT devices often hinder the seamless integration and interpretation of information.

### 2.2   Semantic Interoperability in the IoT

To address the fragmentation of IoT ecosystems, the World Wide Web Consortium (W3C) introduced the Web of Things (WoT) architecture [13], a vendor-neutral framework enabling interoperability among diverse IoT devices. Each IoT device is required to publish a Thing Description (TD), a metadata document that formally defines the device's capabilities, properties, and interactions. Ontologies serve as formal knowledge representation models, utilizing classes, properties, and relationships to organize and describe domain-specific concepts [14]. Within TDs, ontologies are employed to provide a shared semantic model, allowing devices to consistently understand and communicate their data and functionalities. The heterogeneity of data in IoT has led to the development of multiple ontologies aimed at addressing interoperability issues among sensor data. The Semantic Sensor Network (SSN) [15] ontology published as a W3C Recommendation. The Sensor, Observation, Sample, and Actuator (SOSA) [16] ontology is a lightweight version of SSN. The Smart Applications REFerence (SAREF) [17] ontology specifies concepts in the smart appliances domain. However, this approach merely shifts the interoperability challenge to a higher level, as the lack of interoperability persists when IoT components rely on different ontologies [18]. This research aims to develop a system-level semantic layer to facilitate complex data integration, semantic search, and reasoning across heterogeneous IoT datasets.

## 2.3   Digital Twin and Semantic Technologies

Digital Twins (DTs) are receiving considerable focus in the industrial field due to their ability to create a virtual representation of physical assets throughout their lifecycle. This virtual counterpart facilitates advanced decision-making through data analysis, simulation, and machine learning techniques [19]. DTs achieve this by continuously integrating data from various sensors and systems, providing a comprehensive and up-to-date view of industrial operations. This holistic approach enables seamless data integration and analysis, allowing for swift adaptation to dynamic industrial environments. Furthermore, DTs leverage their accurate reflection of the physical environment to model complex interactions and predict potential outcomes, significantly enhancing decision-making processes within the industrial domain [20]. Consequently, DTs bridge the gap between fragmented, siloed solutions and the need for a cohesive and adaptive system. This adaptability ensures the system's continuous evolution alongside the industrial landscape, ultimately promoting sustained efficiency and reliability in managing industrial processes and assets. Traditionally, a DT framework consists of three primary layers: the physical layer, the information layer, and the virtual layer [20,21]. The information layer in a DT typically handles data perception, processing, and analysis. This layer is responsible for collecting data from multiple sources, preparing it for analysis, and applying analytical techniques to extract valuable insights. However, while crucial, this layer often faces challenges in ensuring that data from diverse sources is consistently integrated and interpreted correctly, especially when dealing with heterogeneous data types and formats. A semantic layer, powered by ontologies, offers a standardized and structured approach to representing and integrating data from different sources. Ontologies ensure that data elements are understood and interpreted uniformly, leading to more reliable data analysis. Integrating DTs with semantic technologies enables effective management of heterogeneous data and the seamless merging of fragmented data silos. Therefore, this paper proposes a DT framework that leverages IoT and semantic technologies to accurately mirror industrial environments.

## 3   Proposed Digital Twin Framework for Industrial Working Environment

To provide a digital representation of industrial working environments, this study proposed a three-layer DT architecture, illustrated in Fig. 1. Each layer facilitates bidirectional data transfer. Within the physical layer, multiple edge devices are deployed in the factory to perceive and transmit environmental data. These edge devices serve as the foundational sensory network that captures real-time data from the physical environment. The information layer, utilizing the microservice architecture, is responsible for data processing and knowledge integration. By leveraging microservices, information processing is decomposed into smaller, independent services that manage specific data processing tasks. This approach

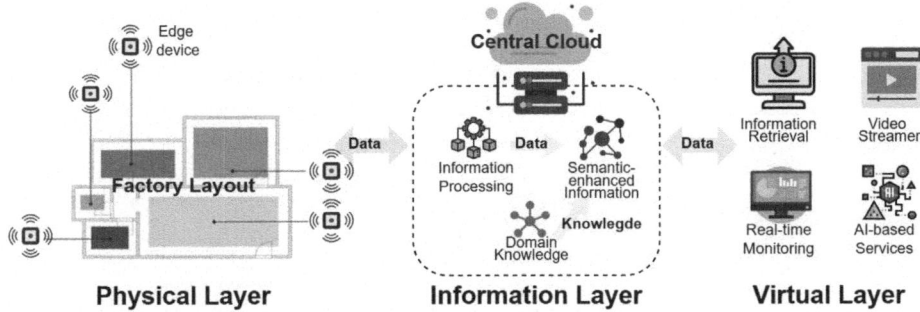

**Fig. 1.** The overview of the proposed Digital Twin for industrial working environments.

facilitates easier maintenance, deployment, and scaling of individual functionalities, which is advantageous in a dynamic industrial environment. Furthermore, this study introduces domain knowledge to describe the perception of edge devices, aiming to integrate heterogeneous data sources collected by these devices at the factory level and ensure harmonization and comprehensibility for further analysis. Through the fusion of semantic-enhanced data with streaming data, the system enables accurate interpretation of complex industrial environments and facilitates the cohesive integration of multiple edge devices to reflect the entire industrial working environment. In the virtual layer, end users can interact with digital services to gain a comprehensive understanding of the working environment. This interaction provides users with insightful visualizations and analytics that reflect the real-time state of the factory, enabling proactive management and optimization of industrial processes. The subsequent sections introduce the proposed domain knowledge model and system architecture of the DT.

### 3.1 Domain Knowledge Model Within the Information Layer

This paper proposes a standardized knowledge representation using ontological models to facilitate the integration of heterogeneous environmental data from various edge devices. This approach consolidates all observed data, offering a holistic description of the working environment. Figure 2 details a conceptual ontological model (prefix: *isien*) designed to represent the edge device's perception in the manufacturing domain. The model leverages established vocabularies and domain knowledge by reusing existing ontologies, including SOSA [16], QUDT [22], GeoSPARQL [23], InPro [14], and BOT [24]. This approach avoids overlapping and redundant modeling and fosters reliability of developing model [25].

The concepts and relationships from SOSA ontology (prefix: *sosa*) are employed to represent sensor-related information for the edge device. A *sosa:Sensor* produces a corresponding *sosa:Observation*, which acts as a procedure to calculate a *sosa:Result* for a specific *sosa:ObservedProperty* of a *FeatureOfInterest*. The *qudt:QuantityValue* class, which is part of the QUDT ontology and prefixed as *qudt*, is a subclass of *sosa:Result* that is used to express values

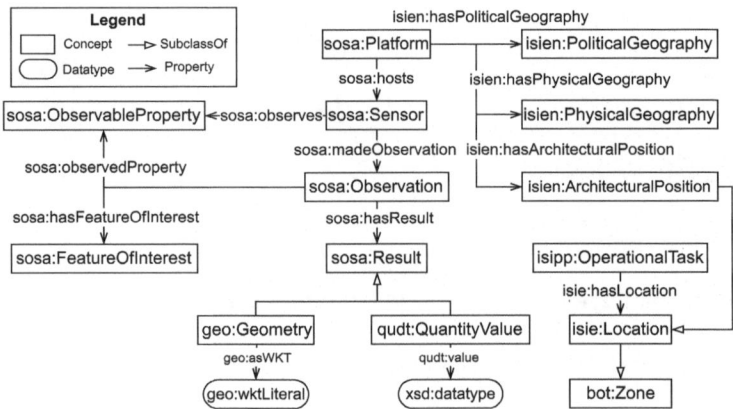

**Fig. 2.** A conceptual ontology for describing edge devices and their observations of industrial working environments.

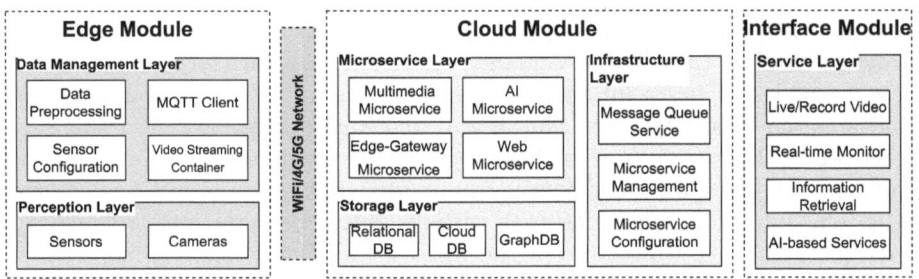

**Fig. 3.** The system architecture of the proposed Digital Twin.

with particular units. In GeoSPARQL, the *geo:Geometry* class, which is prefixed as *geo*, is a subclass of *sosa:Result* and is specifically utilized for representing spatial data. The *isien:PoliticalGeography* class contains information about the political aspects, such as the street, city, and country, related to the edge device. The *PhysicalGeography* class represents the edge device's precise geographical position (e.g., GPS coordinates). The *ArchitecturalPosition* class, a subclass of *isie:Location* from InPro, describes the placement of the edge device within the building's topology. The *isie:Location* class is a subclass of the *bot:Zone* within BOT. The *isipp:OperationalTask* in InPro is connected to the *isie:Location* associated with each production process. Moreover, the proposed ontological model can function as the environmental module for InPro, facilitating the incorporation of environmental information and industrial production processes.

## 3.2   System Architecture of the Proposed Digital Twin

The system architecture of the proposed DT is illustrated in Fig. 3, consisting of three fundamental components: the edge module, the cloud module, and the interface module. The edge module serves as the system's frontline, integrating various IoT sensors for data perception and transmission. This module is structured into two layers: perception and data management. The perception layer acts as an interface between the physical environment and the digital domain, managing hardware components such as sensors and cameras. It is primarily responsible for data acquisition, enabling the retrieval of sensor data and video streams essential for monitoring diverse aspects of industrial operations. The data management layer focuses on processing and organizing collected data. Its key functionalities include data preprocessing (e.g., filtering and cleaning) and data publishing for seamless transmission of processed data to the cloud for further analysis or storage. Additionally, it facilitates sensor configuration, allowing users to customize sensor settings based on specific monitoring needs. Finally, this layer features a video stream container to support video feed transmission to multimedia servers.

To complement the limited computation and storage capacities of the edge module, the cloud module functions as a centralized hub for data processing, storage, and analysis. The cloud server architecture is structured into three distinct layers: the microservice layer, the storage layer, and the infrastructure layer, each addressing specific aspects of microservices-based functionality and data management. The microservice layer contains a set of specialized microservices. The Multimedia Service facilitates live video streaming and video-on-demand functionalities, while the Artificial Intelligence (AI) Service employs machine learning algorithms to analyze data and derive actionable insights. The Edge-Gateway Service serves as the communication interface between the edge module and the cloud module. The Web Service provides web-based visual and interactive UIs for direct user interaction. The storage layer supports diverse database technologies tailored to various data types and usage scenarios. For instance, sensory data is stored in a structured format within a relational database, while cloud databases manage large volumes of video stream data. Graph databases are used for the integration of streaming data and ontological models. Within the cloud module, the infrastructure layer supports the deployment and operation of microservices. The Message Queue Service establishes a real-time data pipeline, facilitating seamless communication and data exchange among different microservices, thereby ensuring efficient coordination and synchronization of distributed processing tasks. Additionally, microservice management and configuration streamline the centralized administration and orchestration of microservices across the entire cloud infrastructure.

The interface module provides a range of services for end users, encompassing live video monitoring, recorded video retrieval, real-time monitoring, information retrieval, AI-enhanced decision-making, and edge device configuration. These functionalities allow stakeholders to oversee the working environment of a factory remotely. Through the interface module, stakeholders obtain insights into the current state of the factory floor during the production processes. Furthermore, this layer facilitates human supervisory control, empowering stakeholders to remotely adjust parameters, initiate workflows, and respond promptly to

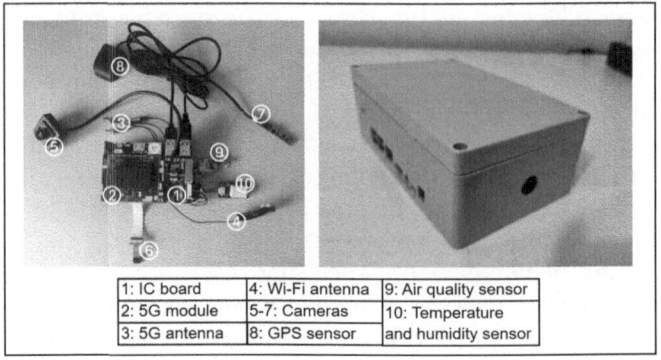

| 1: IC board | 4: Wi-Fi antenna | 9: Air quality sensor |
|---|---|---|
| 2: 5G module | 5-7: Cameras | 10: Temperature |
| 3: 5G antenna | 8: GPS sensor | and humidity sensor |

**Fig. 4.** The design of the edge device: left is the hardware layout of the edge device, and right is the final assembled device.

alerts or emergencies. This capability enhances operational efficiency, ensures adherence to safety regulations, and promotes proactive maintenance strategies.

## 4    System Implementation

### 4.1    Hardware Design of the Edge Device

This study developed a mobile edge device called the Digital Twin Box (DTB) for capturing environmental data, shown in Fig. 4. The DTB features an RK3399 System-on-Chip (SoC) housed on an IC (Integrated Circuit) board, operating on Ubuntu 20.04, supporting edge computing tasks. An industrial-grade Quectel RG500U IoT module integrated with the IC board enables 5G Non-Standalone (NSA) and Standalone (SA) operation, ensuring backward compatibility with 4G/3G networks. For comprehensive visual monitoring, the DTB incorporates three high-definition, wide-angle cameras capturing video streams from various perspectives. Additionally, a GPS sensor provides physical location data, while temperature, humidity, and air quality sensors collect environmental data. The compact design ($20\,cm \times 15\,cm \times 8\,cm$) allows for easy deployment on the factory floor, as depicted in Fig. 4 (right side).

### 4.2    The Knowledge Model of the Edge Device (DTB)

The classes and relationships that describe the DTB (prefix: *dtb*) are illustrated in Fig. 5, based on the proposed domain knowledge model. A *DTB* is a subclass of the *sosa:Platform*, comprising of various sensors, such as *dtb:TemperatureHumiditySensor*, *dtb:AirQualitySensor*, *dtb:GPS*, and *dtb:Camera*. The *dtb:TemperatureHumiditySensor* consists of two distinct observations, *dtb:TempertureObservation* and *dtb:HumidityObservation*. These observations target the *quantitykind:Temperature* and *quantitykind:RelativeHumidity* of the *isie:Location*. The results, *dtb:TemperatureResult* and *dtb:HumidityResult*,

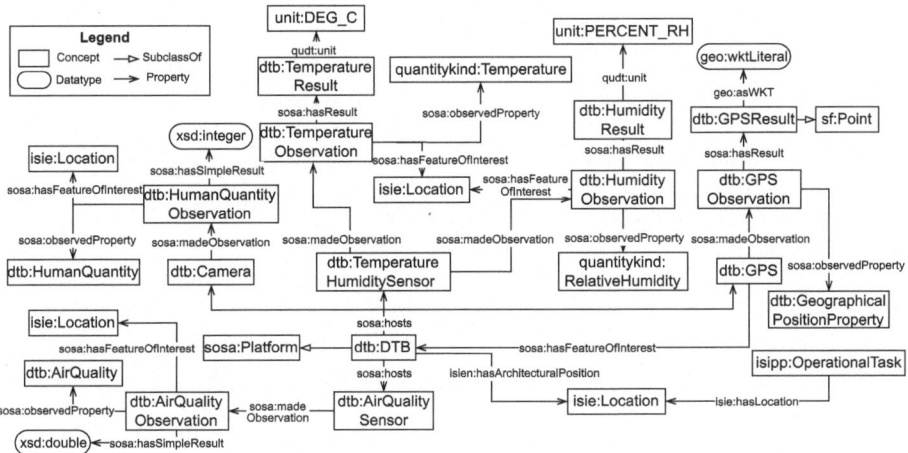

**Fig. 5.** The classes and relationships of the domain ontology pertaining to the designed edge device.

are subclasses of *qudt:QuantityValue*, representing the numerical values with units specified by *unit:DEG_C* and *unit:PERCENT_RH* from the QUDT ontology. Similarly, the *dtb:AirQualityObservation* is designed to measure a numerical value representing the air quality (*dtb:AirQuality*) of the corresponding *isie:Location*. Additionally, the geographical position of the DTB (*dtb:DTB*) is described using the *sf:Point* class from GeoSPARQL, employing the wkTLiteral notation for spatial representation. Finally, leveraging the *isie:Location* class, observations from the DTB can be associated with a specific production process within the industrial environment. Building upon the well-defined semantic structure provided by the ontology model, an instance model is subsequently generated. This instance model facilitates the incorporation of real-time streaming data from the DTBs. The data is then stored in a graph database of the central server, enabling the creation of a comprehensive knowledge graph that furnishes a detailed description of industrial working environments.

### 4.3   The Microservice-Based System Architecture of the Cloud Server

Figure 6 depicts the microservices-based system implementation within the cloud server. As shown in Table 1, various technologies are utilized to construct the cloud server. The microservice layer leverages Docker for containerization to ensure microservice portability and scalability. ZLMediaKit is utilized for providing the Live Video Streaming Service. The Vue.js development framework is employed within the Web Service for the development of dynamic and responsive UI. The EMQX software is utilized to realize the Edge Gateway Service for the MQTT broker. Notably, this work integrates YOLOv8 [26] for object recognition within the AI Service. In the microservice infrastructure layer, NACOS

**Table 1.** The implementation details of the cloud server.

| Layer | Component | Description | Technology |
|---|---|---|---|
| Microservice | Containerization | Microservice portability & scalability | Docker |
| | Multimedia Service | Live video streaming & video-on-demand | ZLMediaKit |
| | AI-based Service | Data analysis with ML/AI algorithms | YOLOv8 |
| | Edge-Gateway Service | Communication bridge (edge-cloud) | EMQX |
| | Web Service | Interactive UI | Vue.js |
| Storage | Relational Database | Stores sensor data | MySQL |
| | Cloud Storage | Stores large video datasets | Native cloud storage solutions |
| | Graph Database | Stores linked data | Neo4j |
| Infrastructure | Container Orchestration | Manages microservices containers | Kubernetes |
| | Message Queue Service | Data pipeline | RabbitMQ |
| | Microservice Configuration & Registration | Discovery & configuration | NACOS |

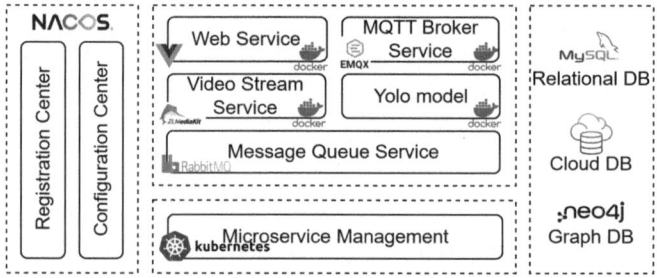

**Fig. 6.** The microservices-based system architecture in the cloud server.

is implemented for microservice configuration and registration discovery, while Kubernetes manages container orchestration for microservices. Furthermore, RabbitMQ is employed as the Message Queue Service to enable asynchronous communication among various microservices within the system. The storage layer utilizes MySQL to store logical relational data, while native cloud storage solutions are leveraged for storing video data. This configuration facilitates swift retrieval and simultaneous storage/playback of extensive video datasets. Neo4j is used as a graph database to implement the instance model of the proposed ontology in the form of knowledge graphs.

## 5   Validation

This study implemented three DTBs within a laboratory-scale manufacturing environment at the Aalto Industrial Internet Campus (AIIC) [27]. Cloud services were provided by a German cloud infrastructure provider. The effectiveness of the proposed DT architecture was evaluated through three key aspects: information retrieval capability, UI usability, and latency performance.

### 5.1   Information Retrieval

In this section, we explore the utility of the proposed DT for governing heterogeneous data sources and fostering sensor interoperability to establish a high-fidelity replica of a physical environment. To assess the effectiveness of this approach, a series of evaluation questions are presented in Table 3. These questions are designed to assess the semantic DT's capacity to retrieve accurate and relevant information, thereby validating its performance within real-world applications. Three DTBs are denoted as 'DTB_1', 'DTB_2', and 'DTB_3'. Each DTB is assigned to a specific zone within the AIIC: 'DTB_1' is assigned to 'Zone_A', 'DTB_2' to 'Zone_B', and 'DTB_3' to 'Zone_C'. Table 2 shows five Cypher queries corresponding to the questions outlined in Table 3. The first query addressed Question 1 by retrieving the number of DTBs within the AIIC. Queries 2 and 3 responded to Questions 2 and 3, which focus on providing the quantitative analysis of sensor data, including the average temperature and maximum humidity. Query 4, aligned with Question 4, identified zones where there are two or more humans. Finally, Query 5 answered Question 5 by calculating the distance between two DTBs using Cypher's spatial functions.

### 5.2   User Interface

This work realizes the interface module as a web UI for end users, providing a comprehensive suite of functionalities, as depicted in Fig. 7. Live video streaming enables the retrieval of real-time video feeds for remote monitoring, along with AI-driven insights. As illustrated in Fig. 7, YOLOv8 is employed to perform real-time people counting through video stream analysis. The GPS sensor embedded into the system allows for viewing the edge device's geographical locations on a map. The video recording feature enables the review of recorded videos. The monitoring service encompasses both real-time and historical data, providing users with a holistic view of the system's operational status. Finally, the management features include device pairing, switching, and sharing, which streamline user control over device setups and operational parameters.

### 5.3   Latency Measurement

To assess the system's performance under varying network connectivity, latency measurements were conducted in three distinct network environments: WiFi

**Table 2.** Cypher statements for the validation questions.

| |
|---|
| #**Query 1** |
| MATCH (:Space{name:"AIIC"}) - [:containsZone] → (loc:Location), |
| (dtb:DTB) - [:hasArchitecturalPosition] → (loc) RETURN count(dtb) |
| #**Query 2** |
| MATCH (:Space{name:"AIIC"}) - [:containsZone] → (loc:Location), |
| (tem:TemperatureObservation) - [:hasFeatureOfInterest] → (loc), |
| (tem) - [:hasResult] → (res:TemperatureResult) RETURN AVG(res.value) |
| #**Query 3** |
| MATCH (h:HumidityObservation) - [:hasFeatureOfInterest] → |
| (loc:Location), (h) - [:hasResult] → (res:HumidityResult) |
| RETURN loc.name AS location, res.value AS humidity |
| ORDER BY res.value DESC LIMIT 1 |
| #**Query 4** |
| MATCH (:Space{name:"AIIC"}) - [:containsZone] → (loc:Location), |
| (hq:HumanQuantityObservation) - [:hasFeatureOfInterest] → (loc) |
| WHERE hq.hasSimpleResult ≥ 2 |
| RETURN loc.name AS location, hq.hasSimpleResult AS value |
| #**Query 5** |
| MATCH (:DTB {name:"DTB_1"}) - [:hosts] → (:GPS) - [:madeObservation] |
| → (:GPSObservation) - [:hasResult] → (res1:Point), |
| (:DTB {name:"DTB_2"}) - [:hosts] → (:GPS) - [:madeObservation] |
| → (:GPSObservation) - [:hasResult] → (res2:Point) |
| RETURN point.distance(res1.value, res2.value) AS distance |

**Table 3.** Specified questions and answers for the information retrieval case.

| Specified questions | Answers |
|---|---|
| 1. How many DTBs are in 'AIIC'? | 3 |
| 2. What is the average temperature of the 'AIIC'? | 20.05 |
| 3. Which location has the maximum humidity? | Zone_B, 56.00% |
| 4. In which AIIC's location are there two people or more? | Zone_A, 2 |
| 5. What is the distance between 'DTB_1' and 'DTB_2'? | 24.20 |

and 4G. Within each network, a dataset of 1000 samples was collected. Each sample consisted of four timestamps capturing: the timestamp when the edge device publishes the message, the timestamp when the central platform receives the message, the timestamp when the central platform sends the message, and the timestamp when the UI receives the message. It's important to note that the two network environments (WiFi and 4G) solely influence the communi-

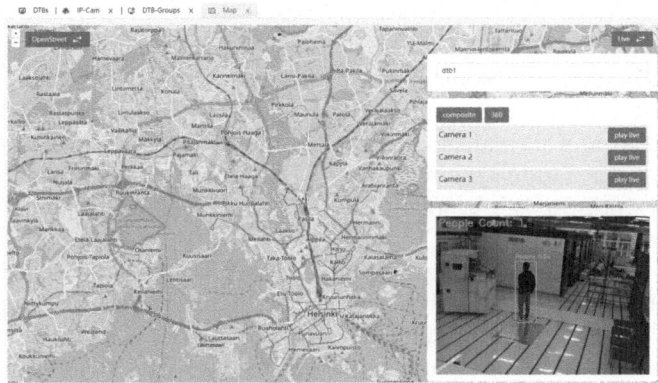

**Fig. 7.** The screenshot of the UI of the proposed DT.

**Table 4.** The latency measurement under WiFi and 4G networks.

| Network | Condition | Mean ($ms$) | Range($ms$) | SD |
|---------|-----------|-------------|-------------|-----|
| WiFi | Edge-UI | 285.451 | 113.390 | 0.021 |
| | Edge-Central | 69.166 | 49.780 | 0.005 |
| 4G | Edge-UI | 413.318 | 650.510 | 0.071 |
| | Edge-Central | 117.263 | 228.110 | 0.040 |

cation between the edge device and the central platform. The communication between the central platform and the UI likely occurs on a separate, potentially local network, and therefore remains independent of the chosen network type. The statistical outcomes of these measurements, including Mean, Range, and Standard Deviation (SD), are presented in Table 4 for each network condition. The table differentiates between "Edge-UI" and "Edge-Central" latency. "Edge-UI" represents the total latency experienced between the edge device and the UI, while "Edge-Central" refers solely to the latency between the edge device and the central platform. The average "Edge-UI" latency measured under WiFi is 285.451 ms, while under 4G it increases to 413.318 ms. Similarly, the "Edge-Central" latency follows the same trend. WiFi demonstrates a lower latency of 69.166 ms compared to 4G's 117.263 ms. Moreover, the average latency for central platform handling was measured to be 11.259 ms, while the average latency between the central platform and the UI is 281.143 ms. In summary, the total latency of our platform is lower than 500 ms for sensor data transmission for the edge device to the end UI.

## 6 Discussion and Limitation

This research aims to provide an efficient approach to perceiving, integrating, and interpreting industrial environmental data, ensuring an adaptive, interoperable, and cohesive digital representation. To achieve this objective, we introduce a semantic-enhanced DT that accurately reflects the complexities of industrial working environments. The main contributions of this work are: 1) Integrating DT with semantic technologies to enhance semantic interoperability; 2) Introducing a unified and formalized ontological model to describe edge devices and their observations of industrial working environments; 3) Applying an edge-cloud system framework for the proposed DT, where the cloud platform leverages a microservices architecture.

First, we integrate DT with a semantic layer that leverages semantic technologies, including ontologies and knowledge graphs (within the graph database Neo4j). The Thing Description (TD) provided by W3C primarily focuses on defining the semantics of individual entities (such as devices and sensors) in IoT environments, offering a framework for describing their properties and behaviors. In our work, we extend beyond individual entities by utilizing knowledge graphs to capture not only their relationships and dependencies but also the interconnected nature of the entire system. This allows for more complex domain knowledge representation and facilitates advanced reasoning and inference capabilities. In the context of DTs, knowledge graphs provide several key benefits. They allow for a unified representation of the properties of individual assets, while also capturing their relationships, dependencies, and interactions with other system components. By using an ontology-based structure, knowledge graphs enhance data integration, making it easier to incorporate and represent data from various sources. Additionally, knowledge graphs enable advanced analytics techniques, such as graph-based analysis, reasoning, and inference, which are invaluable for gaining deeper insights into complex systems and supporting informed decision-making.

Additionally, this paper proposes an ontological model designed to describe edge devices and their perception within the industrial domain. Leveraging established ontologies such as SOSA and QUDT, the model comprehensively represents sensor platforms, their observations, and data structures. It enables the semantic description of various sensors, including temperature, humidity, air quality sensors, GPS devices, and cameras. The model also integrates GeoSPARQL and supports the incorporation of AI models with cameras, enabling capabilities such as human quantity detection. Through integration with InPro, the system establishes connections between sensor data and specific production processes within industrial facilities. Moreover, the semantic model enables seamless connectivity of edge devices at the factory field level, providing a holistic view of the industrial environment. The proposed ontological model facilitates streamlined environmental data management and enhances interoperability across the industrial IoT landscape.

Lastly, this study proposes an edge-cloud system architecture designed to address the limitations of edge devices regarding computational and storage resources. The architecture utilizes a microservices framework within the cloud platform. By decomposing complex applications into four loosely coupled services, the framework promotes flexibility, scalability, and maintainability. The modular nature of microservices enhances fault isolation and resilience, ensuring robust service operation and continuity even during system failures or environmental changes. The Information Layer's microservices offer standardized third-party API interfaces. Stakeholders can customize system logic by generating an application token and accessing relevant functions in the virtual layer. This setup allows large-scale systems to interact directly with the virtual layer through API calls, tailoring functionality to specific needs. Thus, the architecture facilitates efficient utilization of edge resources while leveraging the scalability and flexibility of cloud computing paradigms in industrial IoT applications.

While our research has demonstrated the feasibility of the proposed approach, several limitations need to be addressed. One of the most critical concerns in any IoT environment is data security and privacy, particularly given the system's role in collecting data from industrial settings. The data collected often includes sensitive and proprietary information, making it essential to implement robust privacy safeguards to prevent unauthorized access and ensure compliance with regulations such as the General Data Protection Regulation (GDPR). To enhance data security, one potential solution is to enforce data sovereignty by separating data providers from data service providers. This separation can help minimize risks associated with data breaches and unauthorized access. Additionally, for video streaming, it is imperative to implement de-identification techniques to protect the privacy of human operators. These measures are crucial for maintaining the confidentiality of personal information and upholding privacy standards within industrial IoT applications.

Another limitation pertains to the ontological model developed for our proposed DTB. While the model is tailored to our specific implementation, integrating it into larger, more complex systems of systems requires ontology alignment to facilitate seamless information integration. This process involves harmonizing different ontological frameworks to ensure coherent data exchange and interoperability across diverse systems. Moreover, for systems that lack a semantic representation, additional middleware is necessary to bridge the gap between non-semantic and semantic systems. The need for such middleware introduces complexity and potential performance overhead, which must be carefully managed to prevent degrading system efficiency and responsiveness.

## 7 Conclusion and Future Works

The industrial working environment is becoming increasingly complex. Developing a Digital Twin (DT) of this environment is crucial for smart factories to adequately assess environmental impacts and make more informed decisions. However, existing DT encounters several limitations, such as 1) the lack of an approach to digital representation of industrial working environments, and 2) the heterogeneity of environmental data within the factory field level. Therefore, this study introduces a semantic-enhanced DT tailored for industrial working environments. The system architecture comprises three main modules: edge module, cloud module, and interface module. The edge module supports various environmental data perceptions and transmissions from the field level. The cloud module adopts a microservices approach to ensure scalability and facilitate seamless maintenance and updates. The interface module offers a range of functionalities for end-users, empowering them to effectively monitor and track the working environment. Moreover, the integration of the semantic model enables semantic interoperability and interconnectivity among all edge devices. Validation of the system has demonstrated its feasibility in terms of information retrieval, user interface usability, and performance.

To enhance the capabilities of the proposed platform, future works would contain: 1) Integration of information retrieval capabilities with the web-based UI to provide more intuitive and user-friendly operations and information visualization for end-users. 2) Support for industrial protocols like OPC Unified Architecture (OPC UA) within the system architecture to enhance interoperability and compatibility with industrial equipment and networks. 3) Validate the system's performance in a more complicated industrial working environment.

**Acknowledgement.** This work is partially supported by Business Finland for the co-innovation project through the Necoverse project (Grant 10768/31/2022), the European Union's Horizon Europe program for Research and Innovation through the aerOS project under (Grant No. 101069732), the 6G-SANDBOX project (Grant No. 101096328), and the RIGOUROUS project (Grant No. 101095933). The paper reflects only the authors' views. The Commission is not responsible for any use that may be made of the information it contains.

# References

1. Ustundag, A., Cevikcan, E.: Industry 4.0: Managing the Digital Transformation. Springer, Cham (2017)
2. Soori, M., Arezoo, B., Dastres, R.: Internet of things for smart factories in industry 4.0, a review. Internet Things Cyber-Phys. Syst. **3**, 192–204 (2023)
3. Vaidya, S., Ambad, P., Bhosle, S.: Industry 4.0–a glimpse. Procedia Manuf. **20**, 233–238 (2018)
4. Wang, J., Xu, C., Zhang, J., Zhong, R.: Big data analytics for intelligent manufacturing systems: a review. J. Manuf. Syst. **62**, 738–752 (2022)
5. Teh, H.Y., Kempa-Liehr, A.W., Wang, K.I.-K.: Sensor data quality: a systematic review. J. Big Data **7**(1), 11 (2020)
6. Zheng, X., Lu, J., Kiritsis, D.: The emergence of cognitive digital twin: vision, challenges and opportunities. Int. J. Prod. Res. **60**(24), 7610–7632 (2022)
7. Fürber, C., Fürber, C.: Semantic technologies. In: Data Quality Management with Semantic Technologies. Springer, Wiesbaden (2016). https://doi.org/10.1007/978-3-658-12225-6_4
8. Muralidharan, S., Yoo, B., Ko, H.: Designing a semantic digital twin model for IoT. In: 2020 IEEE International Conference on Consumer Electronics (ICCE), pp. 1–2. IEEE (2020)
9. Yang, C., Lan, S., Wang, L., Shen, W., Huang, G.G.: Big data driven edge-cloud collaboration architecture for cloud manufacturing: a software defined perspective. IEEE Access **8**, 45938–45950 (2020)
10. Hossain, M.L., Abu-Siada, A., Muyeen, S., Hasan, M.M., Rahman, M.M.: Industrial IoT based condition monitoring for wind energy conversion system. CSEE J. Power Energy Syst. **7**(3), 654–664 (2020)

11. Palazon, J.A., Gozalvez, J., Maestre, J.L., Gisbert, J.R.: Wireless solutions for improving health and safety working conditions in industrial environments. In: 2013 IEEE 15th International Conference on e-Health Networking, Applications and Services (Healthcom 2013), pp. 544–548. IEEE (2013)

12. Ahn, J., Park, J., Lee, S.S., Lee, K.-H., Do, H., Ko, J.: SafeFac: video-based smart safety monitoring for preventing industrial work accidents. Expert Syst. Appl. **215**, 119397 (2023)

13. World Wide Web Consortium (W3C): Web of Things (WoT) Architecture, 30 August 2024 (2024). https://w3c.github.io/wot-architecture/

14. Yang, C., et al.: Ontology-based knowledge representation of industrial production workflow. Adv. Eng. Inform. **58**, 102185 (2023)

15. Compton, M., et al.: The SSN ontology of the W3C semantic sensor network incubator group. J. Web Semant. **17**, 25–32 (2012)

16. Janowicz, K., Haller, A., Cox, S.J., Le Phuoc, D., Lefrançois, M.: SOSA: a lightweight ontology for sensors, observations, samples, and actuators. J. Web Semant. **56**, 1–10 (2019)

17. Daniele, L., den Hartog, F., Roes, J.: Created in close interaction with the industry: the smart appliances REFerence (SAREF) ontology. In: Cuel, R., Young, R. (eds.) FOMI 2015. LNBIP, vol. 225, pp. 100–112. Springer, Cham (2015). https://doi.org/10.1007/978-3-319-21545-7_9

18. Novo, O., Francesco, M.D.: Semantic interoperability in the IoT: extending the web of things architecture. ACM Trans. Internet Things **1**(1), 1–25 (2020)

19. Kaur, M.J., Mishra, V.P., Maheshwari, P.: The convergence of digital twin, IoT, and machine learning: transforming data into action. In: Farsi, M., Daneshkhah, A., Hosseinian-Far, A., Jahankhani, H. (eds.) Digital Twin Technologies and Smart Cities. Internet of Things, pp. 3–17. Springer, Cham (2020). https://doi.org/10.1007/978-3-030-18732-3_1

20. VanDerHorn, E., Mahadevan, S.: Digital twin: generalization, characterization and implementation. Decis. Support Syst. **145**, 113524 (2021)

21. Zheng, Y., Yang, S., Cheng, H.: An application framework of digital twin and its case study. J. Ambient Intell. Humaniz. Comput. **10**, 1141–1153 (2019)

22. Rijgersberg, H., Van Assem, M., Top, J.: Ontology of units of measure and related concepts. Semant. Web **4**(1), 3–13 (2013)

23. Battle, R., Kolas, D.: Enabling the geospatial semantic web with parliament and GeoSPARQL. Semant. Web **3**(4), 355–370 (2012)

24. Rasmussen, M.H., Lefrançois, M., Schneider, G.F., Pauwels, P.: BOT: the building topology ontology of the W3C linked building data group. Semant. Web **12**(1), 143–161 (2021)

25. Blanco, C., Lasheras, J., Fernández-Medina, E., Valencia-García, R., Toval, A.: Basis for an integrated security ontology according to a systematic review of existing proposals. Comput. Stand. Interfaces **33**(4), 372–388 (2011)

26. Jocher, G., Chaurasia, A., Qiu, J.: Yolo by ultralytics (2023). https://github.com/ultralytics/ultralytics. Accessed 16 June 2024

27. Yang, C., et al.: Extended reality application framework for a digital-twin-based smart crane. Appl. Sci. **12**(12), 6030 (2022)

# Approaching Interoperability and Data-Related Processing Issues in a Human-Centric Industrial Scenario

Danish Abbas Syed⬤, Walter Quadrini(✉) ⬤, Nima Rahmani Choubeh⬤,
Marta Pinzone⬤, and Sergio Gusmeroli⬤

Department of Management, Economics, and Industrial Engineering, Politecnico di Milano,
Milan, Italy
{danishabbas.syed,walter.quadrini,nima.rahmani,marta.pinzone,
sergio.gusmeroli}@polimi.it

**Abstract.** Industry 4.0 industrial automation paradigm and the related new Operator 4.0 role and pool of competencies are playing a critical role in bringing forth the Digital Transformation to manufacturing industry and SMEs in particular. The human-centric aspect of Industry 4.0 in combination with resilience, sustainability and circularity of manufacturing processes is gaining wider acceptance in Europe and across the globe while the transition towards Industry 5.0 starts to gain momentum as well as the integration of human centric solutions in Industry 4.0 automation systems. The current work uses a three-pronged approach to wearable sensors integrated with existing Industry 4.0 automation systems, by addressing sensor heterogeneity, data interoperability and network latency issues under the umbrella of a single unified and harmonised solution. Such a solution is realised in a realistic industrial scenario showcasing adaptive Human-Robot collaboration and leverages open-source software and open reference architectures.

**Keywords:** Human in the loop · Operator 5.0 · 5G · IIoT · NGSI-LD · CPPS

## 1 Introduction

The so called "Industry 4.0" has gained significant popularity since its introduction in the year 2011. This primary talking point in this new era of manufacturing has been the digitization of industries. With the adoption of this new industrial revolution, concepts like Cyber-Physical Production Systems (CPPS) and Industrial Internet of Things (IIOT) are helping in the transformation of the traditional Factories into "Smart Factories" [1]. Along with the evolution of industries, the role of the operators has also evolved. The term "Operator 4.0" has been used in the context of Industry 4.0 to signify the latest iteration of this evolution [2]. This paradigm calls for the development of "Human-centric Technologies" in a bid to improve the collaboration between humans and technologies/machines. Operator 4.0 typologies like "Healthy Operator", "Super Strength Operator", "Analytical Operator", "Collaborative Operator" can be used to classify terminology for such solutions [3].

© The Author(s) 2025
M. Presser et al. (Eds.): GIECS 2024, CCIS 2328, pp. 21–34, 2025.
https://doi.org/10.1007/978-3-031-78572-6_2

In continuation of Industry 4.0, Industry 5.0 and the corresponding Operator 5.0 can be seen as the next logical evolution. The need for such an evolution was primarily triggered by the COVID-19 pandemic and it exposed several fault lines within the existing systems particularly with respect to the resilience of the production systems. This new phase, while relying on the foundations of the earlier phase, calls for a renewed focus on the operator aspect of the industry [4], also considering that in recent years the attention on the "Social" dimension in the so-called "Triple Bottom Line" [5] has grown in terms of attention [6], in particular following the introduction of the "2030 Agenda for Sustainable Development" [7]. In this sense, the focus on the operators can highly impact these goals, thanks to the technology-assisted approach which can decrease the physical and mental stress of operators, improving their health condition and well-being, and reducing inequalities in accessing certain tasks even for operators with physical inequalities.

With the renewed focus of the industry on the operator, operator-centric technologies and their seamless integration with the existing systems has become of critical importance. One such aspect that the current work deals with is the integration of multivendor, operator-worn sensors in a Human-Robot collaborative workspaces and paced assembly-work scenarios.

The remaining part of the paper is structured as follows: Sect. 2 provides a brief overview of the state of the art and problem description, Sect. 3 briefly describes the testing scenario, Sect. 4 defines the proposed software architecture, Sect. 5 describes the test bed setup, Sect. 6 presents an open-source implementation of the proposed architecture. Section 7 presents the conclusion and future work.

## 2 Problem Description

The aforementioned industrial evolutions bring forth the importance and demand for human-centric solutions which significantly enhance the capabilities and the general wellbeing of the operators at multiple levels. A well-known work [3] made a significant attempt at classifying and formalizing the different typologies to be used for the different categories of operator enhancement. In that context, the current work primarily falls under the banner of "Healthy Operator". Solutions of this kind are aimed at improving the occupational health of the operators (e.g., by reducing fatigue and stress levels). This is achieved by monitoring and processing the biometric signals to evaluate and manage the cognitive and physical stress levels of the operators. To capture the biometric signals, the use of different wearable sensors has been suggested. In this context the use of sensors like Electroencephalograph (EEG), Electromyograph (EMG), Galvanic Skin Response (GSR), Heart Rate (HR), skin temperature, Electrocardiograph (ECG), Heart Rate Variability (HRV), Blood Volume Pulse (BVP), Photoplethysmography (PPG) has been reported in the literature [8]. However, single biometric signal is usually not sufficient to accurately evaluate the stress levels of a person [9] therefore a combination of multiple sensors is commonly used and suggested in literature, in a bid to improve the ability to detect stress [10].

However, connecting multiple sensors for the purpose of developing a stress evaluation system may require a significant integration effort. This is because such a solution demands reliance on multiple sensor manufacturers, and they often rely on different modalities of data transmission and payload formats. The resulting data heterogeneity and data interoperability issues are quite well known and have already been highlighted as one of the challenges for realising Industry 5.0 vision [11]. In this context, efforts to tame the heterogeneity of multivendor devices has already been reported in literature. As an example, a previous work [12] provided an open-source solution integrating multi-vendor Automated Guided Vehicles (AGVs) however, it lacked to address the aspect of cross platform data interoperability.

Up to this point the current section has concentrated on the challenges related to gathering biometric data from multi-vendor wearable devices. However, development of a successful Healthy Operator solution also involves the processing of this data to gain meaningful insights namely the stress or fatigue levels of the operator. Typically, this involves the use of some form of Machine Learning (ML) or Artificial Intelligence (AI) Algorithms [13]. However, running AI or ML workloads often requires significant computational resources which may not be feasible for the CPPS, and there is a need to rely on software and networking architectures which make it feasible to leverage higher computational assets. These computational assets may be available onsite or alternatively through cloud based computational service providers.

Pushing the computational workloads to more computationally capable devices can lead to improvements in the processing capabilities [14] thereby reducing the computational latency. However, it can also introduce significant network related latencies. One solution to this problem that can be seen in the literature is to use fog computation [15]. Similarly slicing of 5G Networks can also be utilised to provide low latencies in critical applications in Industrial settings [16].

The goal of the current work is to provide a solution which addresses the three aspects of wearable sensor integration, which include data interoperability, data heterogeneity, and latency. To address all these points under the umbrella of a single solution, a series of decisions and suggestions related to the selection of software tools, networking, and computational hardware has been used to address each of these issues. Before delving into the intricacies of the proposed solution, a brief discussion about the testing scenario has been provided to set the context of the proposed applications.

## 3  Testing Scenario

To test the developed solution, an experimental scenario has been realized inside a semi-industrial test bed setting. The semi-industrial nature of the test bed has been used here because it facilitates more robust testing and evaluation of the developed solution. This would be more difficult in a real factory because of the potential disruptions to production. The test bed in question hosts two industrial use cases representing two distinct industrial scenarios.

The first use case consists of an independent workstation where an operator equipped with wearable sensors, is supposed to perform a multi-stage repetitive assembly task with adaptive support provided by two collaborative robots. The assembled component in question is composed of 3 parts: a base, a midsection and a threaded top part. The assembly operation consists of following 4 steps:

1. Retrieve assembly components from close by storage areas.
2. Assemble Middle part to the base
3. Assemble the threaded top part
4. Place the finished assembly in the final storage buffer area.

Under the normative conditions steps 2 and 3 are carried out by the operator and collaborative robots are engaged in steps 1 and 4. However when the operator becomes fatigued, the collaborative robots also perform step no 3, relieving him/her from further effort and allowing recovery.

In the second use case the operator works on a workstation which is a part of a paced assembly line for manufacturing of valves. The stations preceding this workstation feed this station at a defined rate and the operator is supposed to keep pace with the feed rate to prevent the line from getting stopped due to pooling up of material at the station input side. The station is fed using two main methods; with an AGV which supplies the main components of the valve assembly at defined intervals, and a set of *gravity flow racks* [17] which feeds standard parts like nuts, bolts etc. The valve assembly consists of 19 sub-components of which 12 are unique. To facilitate fault free assembly, an *Arkite Operator guidance system* [18] is deployed to assist the operator during the complex assembly operation. Under high stress situations, supportive intervention in this scenario is in the form of reduction in the overall takt time of the assembly line.

The overarching goal of both scenarios from an industrial perspective is to make required interventions under high stress or fatigue situations to promote operator well-being, avoiding high risk situations and hence contribute to the prevention of workplace accidents [19].

## 4 Software Architecture

This section describes the proposed software architecture and brief overview of the functional aspects of each of the architectural layers. The overarching goal of the proposed architecture is to present a solution which enables efficient computation by enabling the use of necessary computational resources while also addressing the issues of data interoperability and heterogeneity. The proposed architecture consists of the following main layers:

- Physical layer.
- Gateway layer.
- Middleware and Data persistence layer.
- Service layer.

*The physical layer* consists of an array of industrial assets like wearable sensors, robots, workstations and other industrial assets. These assets rely on a wide array of communication protocols which pose challenges related to integration and interoperability.

*The Gateway layer* consists of an application or a set of applications which are responsible for communicating with the assets in the physical layer. These applications may also include protocol bridge elements which are responsible for translating from the native communication protocol of the sensors to industrial protocols compliant with

RAMI 4.0 specifications which include OPC-UA, MQTT, DDS, AMQP [20] and many others. This consequently helps to addresses the issues related to the heterogeneous data transmission resulting from the use of multi-vendor sensors.

*The Middleware layer* primarily consists of a message broker [21] which stores the context data in a standardised format to promote cross platform interoperability. The layer also allows connecting and interfacing with different industrial protocols to receive northbound sensor data and send southbound messages for actuation of the physical layer elements. Additionally, the layer also hosts one or more types of Databases for different storage requirements.

*The Service layer* consists of analytics, AI/ML algorithms, event processing and visualisation services which can be derived from the data stored in the underlying layer.

With a proper choice of implementation tools, the middleware and the service layers can be flexibly deployed at different computational resources at edge or cloud infrastructure (Fig. 1).

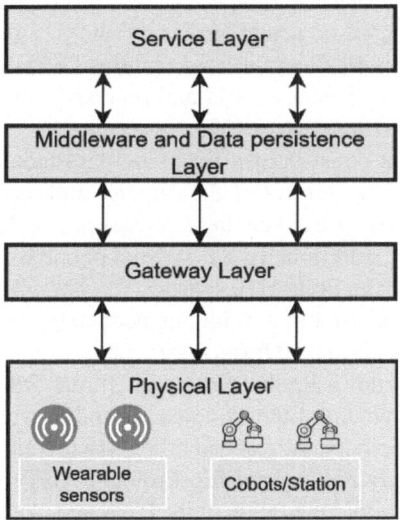

**Fig. 1.** Proposed software Architecture

The proposed software architecture is partly inspired from similar implementations dealing with the integration of industrial assets [22]. However, offers an advancement in terms of added service layer and inclusion of a data persistence to the middle ware layer. Another related implementation of software architecture which specifically deals with human-centric production scenarios is reported in [23]. Although there are some similarities with the current work, however, there are also significant differences. In particular the current work strongly focusses on the use of open-source tools and furthermore solves the interoperability aspects by implementing concepts like linked data. Which is further addressed in Sect. 6 which deals with the implementation of the Architectural model.

## 5   Test Bed Description

The current section provides an in-depth overview of the test bed setup and resources used for validating and testing the proposed solution. In terms of solving identified issues this section primarily focused on addressing the issue of network/computational latency. Figure 2 presents the hardware, and the network diagram used for testing. There are four main elements of the test setup:

- The Shopfloor assets
- The Gateway hardware
- Private 5G system
- An on-premise server

*The Shopfloor* assets consist of a workstation with two collaborative robots and an operator who is equipped with two sets of wearable sensors. 8 surface EMG (sEMG) sensors are placed on major muscles – The bicep, triceps, trapezius and brachioradialis on either side of the body. The use of specific muscle groups is inspired in part from literature references and based on actual trial experiences [24]. Additionally, Polar H10 sensor is used for capturing ECG and related features like HR, R-R interval and HRV. The latter has been selected because ECG and related features are amongst the most widely studied in the industrial context [25].

*The gateway hardware* consists of a set of fan-less industrial PCs which host all the applications related to the control of the robotics and the assembly line station. A windows 10 operating system is used on these devices primarily due to the dependencies of the hosted applications. Additionally, an Android phone is used to host a data logger application for connecting the Polar H10 sensor.

A DELL R740 server is used to provide the necessary computational resources for the resource intensive applications. It hosts the elements of the middleware layer and the service layer. The Server runs a Redhat Enterprise Linux OS and OpenShift container platform is used for deployment of the Middleware and Service layer elements.

For establishing connectivity, the test bed relies on a private 5G network deployment, which relies on 3rd Generation Partnership Project (3GPP) compliant hardware and Open-source software for implementation of the RAN and Core functionalities. To allow non-5G native devices to connect to the access network, Customer Premise Equipment (CPE) is used for relaying the 5G signals to Wi-Fi.

Figure 2 shows how both testbed setup relies on a mixture of both wired, 5G network along with the ability to use Wi-Fi networks. Since the testbed uses an on-premises server with high computational capabilities network and computational latency issues are not expected however given the diversity of connectivity options available it opens the opportunity to make a comparative analysis of network latencies of the three modes. In a more general scenario where on-premises computational capability is limited, the use of cloud services is needed. Under such scenarios, if network latency requirements are quite stringent, Ultra-Reliable Critical Communication service (URCC) public 5G network slice can be utilized. Conversely, Enhanced Mobile Broadband Connectivity slice (eMBB) could be leveraged for high data rates or if the situation demands, a customized network slice could be tailored to cover specific scenarios [26].

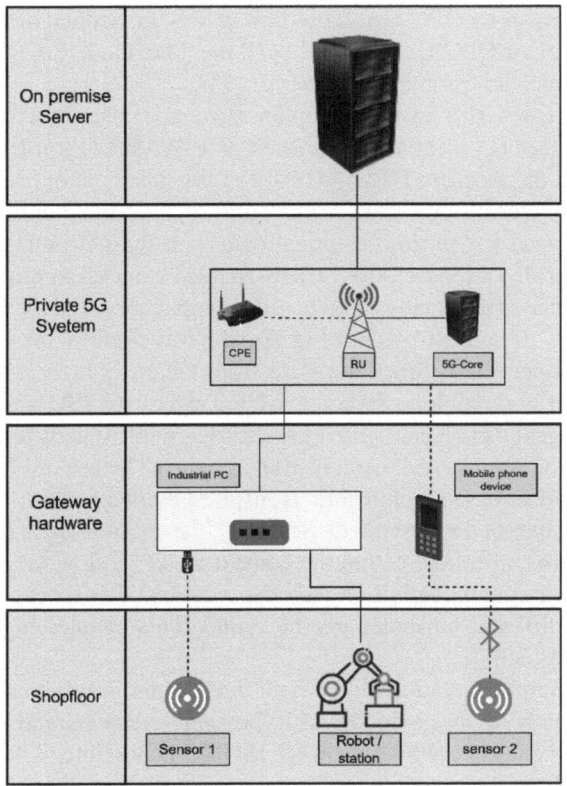

**Fig. 2.** Experimental setup and network architecture

## 6  Software Architecture Implementation

This section provides details about implementation tools used to realize each architectural layer. The implementation relies heavily on the use of open-source implementation tools and tries to satisfy each of the requirements and characteristics described in Sect. 4.

Gateway layer – The gateway layer has been implemented using 3 sets of applications dedicated to each of the hardware elements. For sEMG sensors, a custom application was developed in C++ to connect to the sensors and publish the data to an MQTT Topic. Two programs were developed for collaborative robots which represent two levels of support to the operator. A Node-RED workflow was used to expose an MQTT subscriber client which is used to switch between the two levels of support. Finally, The Polar H10 sensor uses a mobile phone application which publishes data to MQTT topics. As an alternative to the mobile phone app, an additional Python application has also been created to connect the Polar H10 sensor using the Industrial PCs. This has been done to allow a one-to-one comparison of network latencies of the 5G system and the traditional wired system.

Apart from MQTT, OPC-UA protocol was also considered initially, however, MQTT was chosen as a preferred protocol because of its lightweight nature and generally lower latency in comparison to OPC-UA. This however is in neither meant to negate the inherent advantages of OPC-UA protocol [27], nor to exclude the possibility to make these two communication protocols to co-exist [21].

Middleware Layer – This layer was implemented using elements from FIWARE [28] ecosystem. The layer is essentially composed of FIWARE Orion-LD context broker, which is based on the de-facto ETSI-NGSI-LD [28] specification. Mongo-DB is used to hold the latest values of the data in the form of NGSI-LD entities. Additionally, Timescale-DB is used for storing historical data with the help of FIWARE Mintaka. FIWARE Mintaka also exposes APIs which enables external applications to retrieve data from Timescale-DB. Finally, a web server application has been used to hold a static file (namely, "@context", as per Fig. 3) which is used by the Orion-LD context broker for NGSI operations. Since the Data from Gateway layer was transformed to MQTT protocol, it is possible to leverage the IoT agent for JSON from the FIWARE ecosystem. This agent can be configured to subscribe/publish to different MQTT topics corresponding to the sensors and the industrial systems. The use of FIWARE and, more specifically, Orion-LD-based architecture is inspired by two factors, the first one being the open-source nature of the ecosystem. Secondly, the entities which represent different objects like sensors and robots within the context broker follow the JSON-LD format with the addition of context definition which is a set of URIs where detailed semantic definition of the different attributes can be found. This enables high degree of data Interoperability [29, 30].

In the current implementation all the field devices and sensors communicate using MQTT protocol with JSON payloads, with the same elements and minor changes in environmental variables of the IOT agent for JSON it is possible to switch to AMPQ or HTTP protocols. It is also possible to interface with other protocols like OPC-UA, or different payloads like Ultralight 2.0. Alternatively, it is also possible to develop some custom IoT agents in case the protocol/payload combination is not currently supported by the available list of pre-existing solutions.

Service layer – The service layer consists of mainly 2 elements. A Stress detector and a Complex event processing element. Which are both implemented using Python (Fig. 4).

The stress detector element evaluates the operator's stress/fatigue levels based on the historical sensor data queried from the Timescale-DB using the FIWARE Mintaka APIs of the Middleware layer. Our implementation uses sEMG sensor data for fatigue prediction and. Figure 5 shows a pictorial representation of the sequence of steps which are involved in the processing and prediction of Fatigue state of the operator. In essence we are implementing an infinite while loop which carries out the following main steps:

- Data Filtering
- Feature extraction
- Fatigue prediction.

A 5 s initial delay, is introduced at the beginning to ensure that at least 5000 datapoints are available in the Timescale-DB. The data filtration process consists of a 50 Hz notch filter corresponding to AC frequency and 4th order Butterworth band pass filter of 20–450 Hz. Subsequently, 4 frequency domain features namely mean frequency, median frequency, mean power frequency, and zero crossing frequency are extracted from the

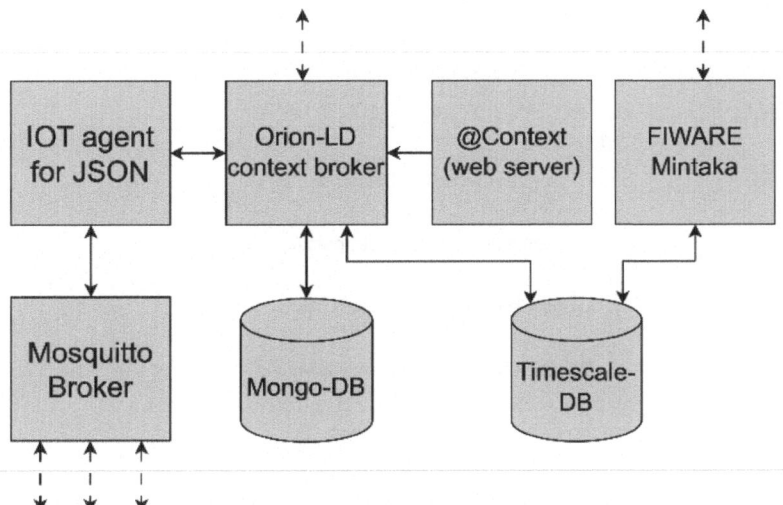

**Fig. 3.** Middleware layer Implementation

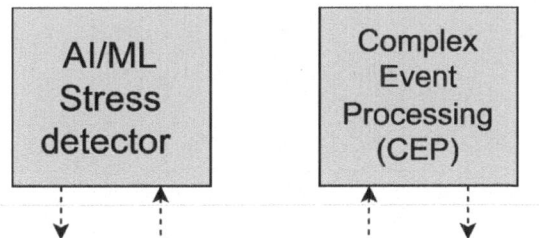

**Fig. 4.** Service Layer Implementation

filtered signal which is used to predict the fatigue state of the different muscles of the operator. The predicted fatigue state is stored in the databases hosted in the Middleware layer through an http PATCH request to the relevant NGSI-LD entity every 1 s. A similar logical workflow has been used to implement the detection of stress using the heart rate sensor. For the sake of simplicity just one of the two is presented here.

The complex event processing directly interacts with the Orion-LD context broker to obtain the latest fatigue and stress state of the operator and based on this an MQTT message sent to the relevant topic is used to switch the operating mode of the robot system. A more detailed logic of the CEP can be seen in Fig. 6. Contrary to the Stress detector logic seen in Fig. 5, an initial delay block is missing in CEP. In order to facilitate this, the NGSI-LD entity holding the fatigue state and stress is always initialized with the lowest values. The alternative use case where production line speed is being changed in response to the operator stress/fatigue state, the downstream communication uses an http POST in place of the MQTT message this is based on the constraints of the system in question.

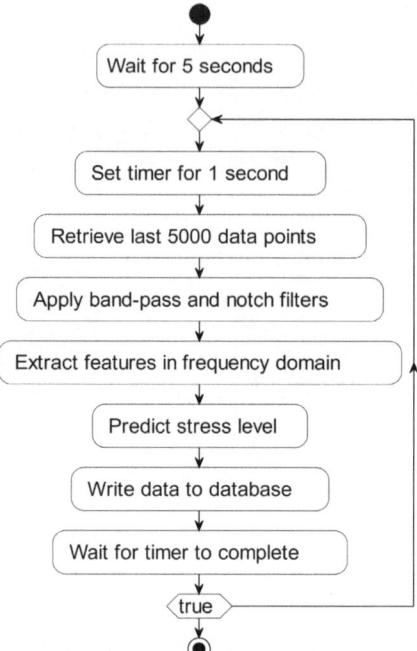

**Fig. 5.** Stress Detector Logic

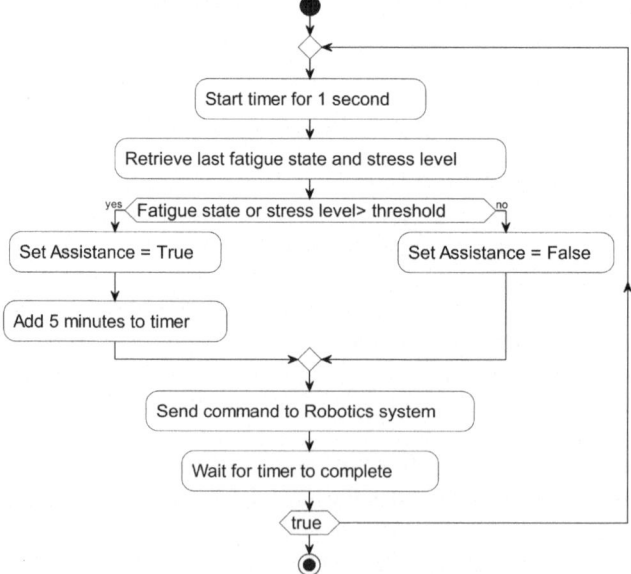

**Fig. 6.** CEP logic

To ensure a user-friendly experience, a simple web interface has also been created which allows the operator to easily start and stop the application. Figure 7 shows some snapshots of the Web interface.

**Fig. 7.** Snapshots from the application web interface

## 7    Conclusion and Future Work

In the recent years the focus of the manufacturing industry has been increasingly shifting towards the adoption of technologies which are directed towards enhancing the well-being of the operator in this direction, the concept of Industry 5.0 has been put forth which essentially renews the focus of the manufacturing world towards the important role of human element in the production sphere. Wearable sensor technology is amongst the key enablers which allows the manufacturing world to advance in the development of human-centric solutions. Integrating these sensors and deriving meaningful and supportive interventions based on the sensor data compels us to deal with the eminent challenges of data interoperability, data heterogeneity and network/computational latencies. To solve the issue of data interoperability and data heterogeneity, the current work outlines an enabling software architecture. And furthermore, takes advantage of the open-source tools like the FIWARE ecosystem to develop some initial implementations of the software architecture. Similarly, the latency issues are managed using a suitable combination of computational resources and the use of low latency 5G network slices.

Although the current works addresses some of the issues related to development and deployment of Human-centric solutions in the industry, solution deployment at an industrial scale still has several challenges. Dealing with biometric data is always associated with General Data Protection Regulation (GDPR) compliance and privacy issues.

This issue has previously been highlighted in literature however, a conclusive solution still eludes us. This is in part due to the inability of the current legislation to properly deal with this issue [31]. As a result, the applications that were developed as a part of the current work primarily focus on a deployment under research settings. The current applications were in fact designed to work in sync with an independent, pre-existing GDPR compliance system that is currently in place in our institution. An extension to the current applications could involve integrating the existing GDPR compliance system into the application itself. On a separate note, while an initial version of the applications has been developed, the applications have seen very limited testing. Therefore, a robust and exhaustive testing needs to be performed to ensure that the applications can be safely deployed. Furthermore, there are several features which remain to be integrated. This as an example includes the addition of role-based access and on a lighter note, a more elegant interface could also be a positive addition. Also, as an extension to the current scope of work the gateway layer could be implemented to intentionally use more than one communication protocol at the same time in a bid to introduce more diversity to the testing scenario and to compare the advantages and disadvantages of using different protocols.

From a purely development perspective, the primary goal of the current stage of work has been to create a working platform using open-source tools. Looking from the perspective of human-centric manufacturing, the current implementation only integrates 2 wearable sensors. Additional sensors like EEG, GSR, Skin temperature etc. could greatly improve the utility of the platform for its potential use by a wider audience. Extending on the same line of thought, creation of Open-source libraries for processing bio signals is an important research tool that can potentially reduce the implementation time and effort of potential adopters. As such it is intended to publish an extended form of the currently developed processing tools in the form of python libraries.

**Acknowledgements.** This work has been supported by the projects INtellegent COlaborative DEployements, Autonomous, scalablE, tRustworthy, intelligent European meta Operating System for the IoT edge-cloud continuum (aerOS), and European Lighthouse to Manifest Trustworthy and Green AI (ENFIELD) which have respectively received funding from the European Union's Horizon Europe research and innovation program under grant agreements 101093069, 101069732, and 101120657, as well as by the HumanTech Project. The HumanTech Project is financed by the Italian Ministry of University and Research (MUR) for the 2023–2027 period as part of the ministerial initiative "Departments of Excellence" (L. 232/2016). The initiative rewards departments that stand out for the quality of the research produced and funds specific development projects.

# References

1. Xu, X., Lu, Y., Vogel-Heuser, B., Wang, L.: Industry 4.0 and industry 5.0—Inception, conception and perception. J. Manuf. Syst. **61**(October), 530–535 (2021). https://doi.org/10.1016/j.jmsy.2021.10.006
2. Romero, D., Bernus, P., Noran, O., Stahre, J., Fast-Berglund, Å.: The operator 4.0: human cyber-physical systems & adaptive automation towards human-automation symbiosis work systems. In: Nääs, I., et al. (eds.) APMS 2016. IFIPAICT, vol. 488, pp. 677–686. Springer, Cham (2016). https://doi.org/10.1007/978-3-319-51133-7_80

3. David, R., et al.: Towards an operator 4.0 typology: a human-centric perspective on the fourth industrial revolution technologies (2016).https://api.semanticscholar.org/CorpusID: 125890216

4. Romero, D., Stahre, J.: Towards the resilient operator 5.0: the future of work in smart resilient manufacturing systems. Proc. CIRP **104**(March), 1089–1094 (2021). https://doi.org/10.1016/j.procir.2021.11.183

5. Elkington, J.: Towards the sustainable corporation: win-win-win business strategies for sustainable development. Calif. Manag. Rev. **36**(2), 90–100 (1994). https://doi.org/10.2307/411 65746

6. Gladysz, B., et al.: Sustainability and Industry 4.0 in the packaging and printing industry: a diagnostic survey in Poland. Eng. Manag. Prod. Serv. **16**(2), 51–67 (2024). https://doi.org/10.2478/emj-2024-0013

7. UN: The Sustainable Development Agenda. https://www.un.org/sustainabledevelopment/development-agenda/

8. Arsalan, A., Majid, M., Nizami, I.F., Manzoor, W., Anwar, S.M., Ryu, J.: Human Stress Assessment: A Comprehensive Review of Methods Using Wearable Sensors and Non-wearable Techniques, pp. 1–76 (2022)

9. Arza, A., et al.: Measuring acute stress response through physiological signals: towards a quantitative assessment of stress. Med. Biol. Eng. Comput. **57**(1), 271–287 (2019). https://doi.org/10.1007/s11517-018-1879-z

10. Naegelin, M., et al.: An interpretable machine learning approach to multimodal stress detection in a simulated office environment. J. Biomed. Inform. **139**, 104299 (2023). https://doi.org/10.1016/j.jbi.2023.104299

11. Taj, I., Jhanjhi, N.Z.: Towards industrial revolution 5.0 and explainable artificial intelligence: challenges and opportunities. Int. J. Comput. Digit. Syst. **12**(1), 285–310 (2022). https://doi.org/10.12785/ijcds/120124

12. Quadrini, W., Negri, E., Fumagalli, L.: Open interfaces for connecting automated guided vehicles to a fleet management system. Proc. Manuf. 406–413 (2020). https://doi.org/10.1016/j.promfg.2020.02.055

13. Nucera, D.D., Quadrini, W., Fumagalli, L., Scipioni, M.P.: Data-Driven State Detection for an asset working at heterogenous regimens. IFAC-PapersOnLine **54**(1), 1248–1253 (2021). https://doi.org/10.1016/j.ifacol.2021.08.149

14. Opara, J., Sahandi, R., Alkhalil, A.: Cloud computing from SMEs perspective: a survey-based investigation (2013)

15. Sarabia-Jácome, D., Usach, R., Palau, C.E., Esteve, M.: Highly-efficient fog-based deep learning AAL fall detection system. Internet Things (Netherlands) **11**, 100185 (2020). https://doi.org/10.1016/j.iot.2020.100185

16. Kalør, A.E., Guillaume, R., Nielsen, J.J., Mueller, A., Popovski, P.: Network Slicing for Ultra-Reliable Low Latency Communication in Industry 4.0 Scenarios, pp. 1–11 (2017)

17. Halim, N.H.A., Jaffar, A., Yusoff, N., Adnan, A.N.: Gravity Flow Rack's material handling system for Just-in-Time (JIT) production. Proc. Eng. 1714–1720 (2012). https://doi.org/10.1016/j.proeng.2012.07.373

18. Bosch, T., Van Rhijn, G., Krause, F., Könemann, R., Wilschut, E.S., De Looze, M.: Spatial augmented reality: a tool for operator guidance and training evaluated in five industrial case studies. In: ACM International Conference Proceeding Series, Association for Computing Machinery, pp. 296–302, June 2020. https://doi.org/10.1145/3389189.3397975

19. Steffy, B.D., Jones, J.W., Murphy, L.R., Kunz, L.: A demonstration of the impact of stress abatement programs on reducing employees' accidents and their costs. Am. J. Health Promot. **1**(2), 25–32 (1986). https://doi.org/10.4278/0890-1171-1.2.25

20. Marcon, P., et al.: Communication technology for industry 4.0. In: 2017 Progress in Electromagnetics Research Symposium - Spring (PIERS), pp. 1694–1697 (2017). https://doi.org/10.1109/PIERS.2017.8262021
21. Quadrini, W., Galparoli, S., Nucera, D.D., Fumagalli, L., Negri, E.: Architecture for data acquisition in research and teaching laboratories. Proc. Comput. Sci. **180**, 833–842 (2021). https://doi.org/10.1016/j.procs.2021.01.333
22. Salis, A., Marguglio, A., De Luca, G., Razzetti, S., Quadrini, W., Gusmeroli, S.: An edge-cloud based reference architecture to support cognitive solutions in process industry. Proc. Comput. Sci. **217**, 20–30 (2023). https://doi.org/10.1016/j.procs.2022.12.198
23. Bettoni, A., et al.: Mutualistic and adaptive human-machine collaboration based on machine learning in an injection moulding manufacturing line. Proc. CIRP 395–400 (2020). https://doi.org/10.1016/j.procir.2020.04.119
24. Giannakakis, G., Grigoriadis, D., Giannakaki, K., Simantiraki, O., Roniotis, A., Tsiknakis, M.: Review on psychological stress detection using biosignals. IEEE Trans. Affect. Comput. **13**(1), 440–460 (2022). https://doi.org/10.1109/TAFFC.2019.2927337
25. Blandino, G.: How to measure stress in smart and intelligent manufacturing systems: a systematic review. Systems **11**(4), 167 (2023). https://doi.org/10.3390/systems11040167
26. Zhang, S.: An overview of network slicing for 5G. IEEE Wirel. Commun. **26**(3), 111–117 (2019). https://doi.org/10.1109/MWC.2019.1800234
27. Silveira Rocha, M., Serpa Sestito, G., Luis Dias, A., Celso Turcato, A., Brandao, D.: Performance comparison between OPC UA and MQTT for data exchange. In: Proceedings of the 2018 Workshop on Metrology for Industry 4.0 and IoT, MetroInd 4.0 and IoT 2018, pp. 175–179 (2018). https://doi.org/10.1109/METROI4.2018.8428342
28. FIWARE Foundation. https://www.fiware.org/catalogue/. Accessed 13 Feb 2024
29. Kumar, S., Jeong, S., Kim, S.: Things data interoperability: linking annotated entities in oneM2M. In: 2022 IEEE 8th World Forum on Internet of Things (WF-IoT), pp. 1–6, IEEE, October 2022. https://doi.org/10.1109/WF-IoT54382.2022.10152120
30. Tolcha, Y., et al.: Towards interoperability of entity-based and event-based IoT platforms: the case of NGSI and EPCIS standards. IEEE Access **9**, 49868–49880 (2021). https://doi.org/10.1109/ACCESS.2021.3069194
31. Cutrona, V., Bonomi, N., Montini, E., Ruppert, T., Delinavelli, G., Pedrazzoli, P.: Extending factory digital twins through human characterisation in Asset Administration Shell. Int. J. Comput. Integr. Manuf. **00**(00), 1–18 (2023). https://doi.org/10.1080/0951192X.2023.2278108

# Leveraging the Industrial Internet of Things (IIoT) for Real-Time CO2 Monitoring, Measurement and Visualization: Technologies, Applications and Future Directions

Mads S-F. Christensen[✉]

BTECH, Aarhus University, 7400 Herning, Denmark
msfchrs@gmail.com

**Abstract.** Global CO2 emissions reduction requires industries to manage and understand their CO2 emission levels in real-time. This paper examines the Industrial Internet of Things (IIoT) for real-time monitoring, measurement, and visualization of reducing CO2 emissions in industrial and environmental domains.

**Methodology:** The methodology consists of a literature review based on peer-reviewed publications and use cases to explore the current state and practical implications. Furthermore, a technical analysis of IIoT systems, CO2 sensors, and data processing techniques is also identified.

**Results:** IIoT systems can support CO2 emission monitoring and accuracy optimization in industrial domains by combining CO2 sensors, wireless communication, and data fusion techniques. In addition, machine learning and artificial intelligence can be used to reduce anomalies in CO2 sensor readings and predictive maintenance of systems.

**Challenges:** Challenges include interoperability, data security and system scalability. To resolve these issues standardized communication protocols, data security methods and implementation barriers should be improved.

**Future Directions:** To enhance data processing and security features, future work should focus on integrating edge computing, artificial intelligence, machine learning, and blockchain techniques. In addition, data visualizations and cost-effective solutions should also be in focus, to provide more adoptable IIoT systems in industrial domains.

**Conclusion:** As IIoT systems and CO2 sensor technologies evolve, IIoT systems can contribute significantly to increasing global air quality and CO2 emission control in industry, agricultural, and urban areas.

**Keywords:** Industrial Internet of Things (IIoT) · CO2 Monitoring · Real-Time Data Processing · Carbon Footprint Monitoring · Air Quality · Security and Privacy in IIoT · Emission Management · Predictive Maintenance · Edge Computing · Machine Learning · Blockchain Technology · Smart Buildings · Smart Cities · Environmental Sustainability

M. Presser et al. (Eds.): GIECS 2024, CCIS 2328, pp. 35–59, 2025.
https://doi.org/10.1007/978-3-031-78572-6_3

# 1   Introduction

The Industrial Internet of Things (IIoT) integrates Internet of Things (IoT) technologies into manufacturing environments. IIoT systems can collect and analyze air quality and $CO_2$ emissions levels to improve sustainability, productivity, and reliability in industrial sectors [1, 2]. Sensor devices are the main components of the IIoT systems architecture, adopting Industry 4.0 and Industry 5.0 practices by leveraging real-time communication among machines [2]. Also, with the integration of edge and cloud computing, secure data management, machine learning (ML) and artificial intelligence (AI), IIoT systems can advance real-time emission monitoring and sustainable practices [2, 3]. Additionally, wireless technologies like LoRa and WAN can enable wireless and real-time $CO_2$ monitoring across domains [4, 5]. Also, incorporating IIoT systems with user visualization tools can improve accessibility to $CO_2$ emissions data for environmental management within industries, manufacturing, and operations [6–9]. Furthermore, monitoring $CO_2$ emissions levels can be essential for enhancing sustainability practices and complying with EU regulations, as industries not complying with the regulations can be fined [10]. High $CO_2$ levels may also indicate inefficient ventilation, which poses health and productivity risks to employees. In addition, by combining IIoT systems and visualization tools, raw $CO_2$ emission data can be converted into digital formats, facilitating decision-making on environmental $CO_2$ emission concerns, and decreasing $CO_2$ emission levels from industrial activities [4, 9, 11, 12].

## 1.1   Research Objectives

This paper examines the benefits and challenges of IIoT systems for real-time $CO_2$ monitoring, measurement, and visualization in industrial environments and makes the following contributions:

- **Technological Framework:** What are IIoT systems, and how can IIoT systems be utilized to monitor $CO_2$ emissions?
- **Practical Examples:** Use cases utilizing IIoT systems for $CO_2$ monitoring or other environmental emissions in real-life scenarios, sectors, and domains.
- **Challenges and solutions:** What are the challenges and system solutions for implementing IIoT systems for $CO_2$ monitoring?
- **Future Directions:** What trends and technologies will support IIoT systems for industrial $CO_2$ monitoring in the future?

By covering these topics, the findings aim to present a thorough overview of how IIoT systems can be utilized to improve industrial sustainability practices and on a global scale.

# 2   Methodology

To analyze the implementation benefits and challenges of IIoT systems, the methodology involves a literature review, technical analysis, and practical use cases. The literature is founded on proceedings from databases, such as IEEE Xplore, ScienceDirect and

Google Scholar, to gather diverse information, including theoretical articles, use cases, and practical applications. Defined criteria were used to screen and evaluate the gathered literature based on relevance to the research objectives and subjected to the following quality assessment:

### 2.1 Inclusion Criteria for Selected Literature

The literature was screened based on the following:

- Peer-reviewed articles and conference proceedings.
- Real-time data processing and environmental emission monitoring using IIoT systems and sensor devices.
- Relevant to industries, manufacturing, agriculture, cities, and buildings.
- Various measurement methods to capture different environmental conditions.

### 2.2 Inclusion Criteria for Selected Use Cases

Real-world use cases were selected based on real-world applications, implementation challenges, and quantifiable benefits. Each use case was reviewed to assess the following:

- Use cases focusing on $CO_2$ emissions or other air pollutants being reduced.
- Scalability of the presented IIoT system technologies.
- Economic viability of IIoT systems at scale.

### 2.3 Analyzing the Literature

The literature was categorized into common themes and topics, and triangulation from multiple sources was utilized to reduce bias in the data. Also, to gain a deeper understanding of the benefits and limitations of IIoT systems and $CO_2$ sensors, a technical analysis was conducted to compare IIoT system performance and practical implementations. This included reliability of data processing speeds, security aspects, and system accuracy in detecting $CO_2$ emissions and anomalies.

## 3   Background and Literature Review

### 3.1 Overview of IIoT System Technologies

Industrial Internet of Things (IIoT) technologies connect sensors, data analysis and wireless communication to enhance industrial processes and operations. [1, 2]. Furthermore, real-time monitoring can be crucial to maintaining efficiency, productivity, and reliability in $CO_2$ emission practices in industries [13, 14]. Correspondingly, data-centric systems form the backbone of Industry 4.0 and 5.0, facilitating communication and cooperation among devices in intelligent industries [15]. IIoT systems encompass sensors for gathering emission data, as well as microcontrollers, edge and cloud computing for data processing, storage, and real-time decision-making, supporting practices and sustainable efficiency in monitoring $CO_2$ emission levels across industries and domains [16, 17].

## 3.2 Existing CO2 Measurement Techniques

Existing techniques for measuring CO2 emissions have advanced due to the emergence of sensor devices like non-dispersive infrared (NDIR) sensors, recognized for their precision and dependability in monitoring CO2 levels [18, 19]. Recently, inexpensive sensors have also been combined with microcontrollers, such as Arduino and Raspberry Pi microcontrollers, making it possible to deliver real-time emission, and air quality monitoring systems, varying from the size of a credit card and to even smaller devices [20, 21]. Additionally, these lightweight IIoT systems can measure emission parameters and other air pollutants at low cost [21]. For example, an Arduino microcontroller, connected to sensors, as depicted in Fig. 1, can provide real-time monitoring of temperature, humidity, and CO2 emissions, making these small IIoT systems useful for various applications, while remaining cost-effective and affordable [22].

**Fig. 1.** Arduino ESP32 and ESP8266 Development Boards with DHT11 and MQ-135 Sensor.

## 3.3 Use Cases and Previous Studies

Research and practical applications have demonstrated that the IIoT systems are capable of monitoring, measuring and visualizing CO2 emissions levels in various scenarios. Implementing IIoT systems in industrial sectors has led to enhancements in monitoring practices, resource management, productivity, and sustainability efforts [9]. In the study [7], an IIoT system utilized by an Arduino microcontroller was linked with a MQ-135 gas sensor for CO2 emission detection. This system also triggers an alarm when emission levels increase and notifies systems authorities. Another project [20] employed an Arduino-based IIoT system for educational purposes by gathering CO2 emission data in different environments and storing the emission data on an integrated SD card. Furthermore, the study [23] shows how edge computing can improve real-time data processing by reducing system latency and monitoring CO2 emissions in real-time. Additionally, [24] utilizes ML algorithms to identify deviations in CO2 levels and strengthen the sensor's overall accuracy.

# 4   IIoT System Architecture for CO2 Monitoring

## 4.1   Components of IIoT Systems

The IIoT system architecture for CO2 monitoring, measurement and visualization consists of a mix of components [1, 16, 17]. Key IIoT system components include various types of CO2 sensors for data collection, transmission mechanisms, and processing techniques for data analysis and visualization.

## 4.2   CO2 Sensor Types for IIoT Systems

Sensors are the building blocks of IIoT systems that detect and measure environmental parameters like CO2 emission levels. However, despite various CO2 sensor types, the sensors have different advantages, like real-time monitoring capabilities, accuracy, and response time. Also, integrating sensors into IIoT system architecture allows for remote monitoring and data availability, which improves CO2 monitoring efficiency. Using advanced CO2 sensor technologies with IIoT system microcontrollers enables continuous monitoring of CO2 levels and timely intervention to maintain sustainable industrial environments [25, 26]. Non-dispersive infrared (NDIR) sensors are utilized for their accuracy and reliability in monitoring CO2 emissions [18, 27]. In addition, capacitive sensors with humidity interference suppression are used for reliable air quality monitoring. Capacitive sensors can also be integrated into portable devices for flexibility and ease of deployment in different environments [28]. Other sensor technologies have also been developed for CO2 monitoring, such as TDLAS sensors [29], solid-state sensors [30] and electrochemical-based sensors [31]. Each CO2 sensor type has its advantages and challenges for industrial and environmental CO2 monitoring, measurement, and visualization in IIoT system applications, described further in the following:

### 4.2.1   Non-Dispersive Infrared (NDIR) Sensors

NDIR sensors measure CO2 by detecting the amount of infrared light CO2 molecules absorb, making them suitable for CO2 measurement and monitoring. NDIR sensors have high emission sensitivity, minimal cross-sensitivity to other gases, and robustness under different environmental conditions, useful for industrial low-cost monitoring applications [27]. Also, NDIR sensors offer low power consumption, fast stabilization time, and wireless deployment compatible with IIoT systems. However, despite their widespread use, NDIR sensors also have limitations, including higher costs and larger sizes, which can hinder integration into more compact or cost-sensitive applications [18].

### 4.2.2   Capacitive Sensors

Capacitive sensors measure CO2 concentrations by detecting capacitance changes. They are known for their low power consumption, fast response time, and can be designed to reduce humidity interference, which is crucial for maintaining accuracy in air quality monitoring systems [28]. Furthermore, capacitive sensors can serve well in environments requiring long-term stability and minimal maintenance [32, 33].

### 4.2.3  Photoacoustic Sensors

Photoacoustic sensors utilize laser light to stimulate CO2 molecules causing them to heat up and expand, resulting in the production of sound waves. These waves are analyzed to determine the composition of a sample using acoustic patterns. This approach is reliable and efficient in most conditions, making these sensors valuable for accurate CO2 monitoring purposes, where precision is crucial [34].

### 4.2.4  Solid-State Sensors

Solid-state sensors detect CO2 through electrical resistance changes when CO2 interacts with the sensor material. These sensors are compact, cost-effective, and can be used in a range of applications, from portable CO2 monitors to fixed installations in buildings. Solid-state sensors offer decent sensitivity and response time, making them versatile for CO2 monitoring [35]. Solid-state sensors can also effectively monitor indoor air quality without significant interference from O2 concentration [30].

### 4.2.5  Tunable Diode Laser Absorption Spectroscopy Sensors

Tunable Diode Laser Absorption Spectroscopy (TDLAS) sensors are known for their unique accuracy and sensitivity in detecting gases like CO2 emissions [36]. TDLAS sensors can provide real-time monitoring with fast response times and operate in harsh industrial environments. Also, TDLAS sensors can measure low gas concentrations with high precision, making TDLAS sensors optimal for critical IIoT systems where reliable and accurate CO2 monitoring is needed [37].

### 4.2.6  Fiber Optic Sensors

Fiber optic sensors are valued for their immunity to electromagnetic interference, high sensitivity, and precision, making them useful for detecting low concentrations of gases like CO2 emissions [38]. Fiber optic sensors can operate in severe environments, such as high temperatures and acidic atmospheres. In addition, fiber optic sensors allow for long-range measurement with minimal signal loss and their small sensor size, makes them suitable for various and challenging IIoT system applications [39].

### 4.2.7  Metal-Oxide Semiconductor Sensors

Metal-oxide semiconductor (MOS) sensors can detect various gases, including CO2, with decent sensitivity and fast response times [40]. Their small sensor size and low power consumption suit well for portable and distributed system applications. Despite limitations in sensitivity and stability compared to other CO2 sensors, MOS sensors can still be a practical choice for various industrial applications [41].

### 4.2.8  Nanoelectronic Sensors

Nanoelectronic sensors can detect CO2 emissions based on a polymer coated NTNFET architecture. The sensor is small, has low power consumption, and can detect CO2 emissions at low temperatures, making them suitable for wireless detection in industrial, but also medical CO2 sensor devices [42].

### 4.2.9  Electrochemical Sensors

Electrochemical sensors detect CO2 by producing electrical signals through a chemical reaction. These signals are proportional to CO2 emission concentration and can be useful for system applications, such as medical and laboratory equipment. However, although electrochemical sensors are not widely used, they are valued for their low detection limits and small size, making these sensors suitable for detailed environmental monitoring [31].

### 4.2.10  Thermal Conductivity Sensors

Thermal conductivity sensors can operate with different gases to conduct heat at different rates. By measuring heat from a heated element and the presence of a gas, CO2 concentration can be inferred based on its thermal conductivity relative to a reference, usually air. Also, thermal conductivity sensors can operate under harsh conditions, making them suitable for industrial applications where sensor robustness is needed [43].

## 4.3  Data Processing

Another consideration when selecting CO2 sensors to be integrated with IIoT systems is how to process the IIoT system data. Edge computing processes system data near the system source, while cloud computing sends and processes system data on remote servers [23, 44]. Edge and cloud computing are both useful for data processing requirements but deliver different benefits and challenges.

### 4.3.1  Edge Computing

Edge computing plays a significant role in minimizing system latency and real-time data processing for IIoT system applications that demand instant data responses [17, 23]. Users can experience fast data feedback by delegating data processing directly to local edge computing servers, as being closer to the data source reduces delays [45, 46]. Furthermore, edge computing can cut system latency and decrease data transfer rates significantly, compared to cloud services, surpassing centralized cloud computing architectures [47–49]. Nevertheless, edge computing also comes with potential risks, such as data processing capabilities and limited storage options. These reduced capabilities may pose difficulties for complex data analyses, requiring substantial computational resources [17, 46].

### 4.3.2  Cloud Computing

Cloud computing is suited to tasks that demand high data resources and scalability [44, 50]. Cloud computing infrastructure is key to resourceful and flexible operations across IIoT system locations, which do not require real-time feedback [51, 52]. Furthermore, cloud computing provides end-users and industries with an array of extensible services by managing resources and extending data services to the edge of networks [50]. However, since the data information needs to travel from the IIoT system origin to external servers, cloud computing also poses challenges and potential risks, with increased delays and latency in the data streams. This aspect is particularly critical for IIoT system applications, which require prompt responses and real-time data processing [53]. Furthermore, IIoT systems relying on cloud services also face security and privacy risks, as data transmission could potentially be intercepted during transfer [54] (Fig. 2).

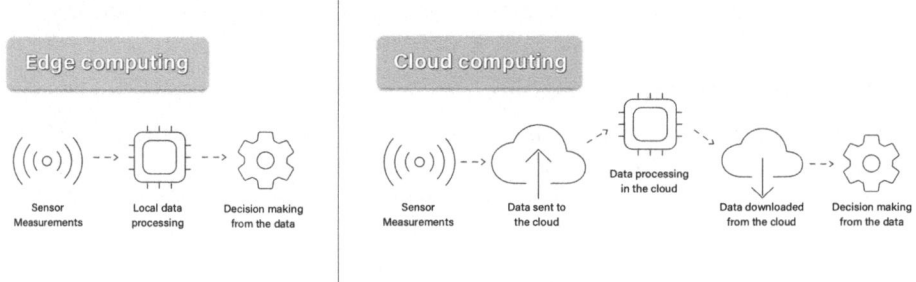

**Fig. 2.** Edge and Cloud Computing.

### 4.3.3  Data Filtering

Another part of the data processing stage is to filter the raw system data, to remove noise and other unwanted information from the IIoT system and sensor data [55]. Noise in the data can derive from several sources, such as sensor malfunction, environmental interference, or transmission errors within the IIoT system. Filters such as resampling filters, low pass filters, median filters, and more advanced statistical methods can help smooth out the IIoT system data, which can be crucial for IIoT systems and connected sensors to accurately measure $CO_2$ emissions [55].

### 4.3.4  Data Anomaly Detection

Anomaly detection is also a vital part of IIoT systems for $CO_2$ monitoring, as it supports identifying unusual data patterns that may indicate malfunction or environmental exposure. ML techniques such as Long-Short-Term Memory (LSTM) networks, and Bayesian and Gaussian processes can be applied to detect system anomalies based on identifying deviations from normal patterns or support by triggering alerts for preventive maintenance or actions based on the collected data from the IIoT system sensors [24, 55].

### 4.3.5  Data Calibration

Calibration is also key to IIoT systems and sensors that measure $CO_2$ emission levels. As IIoT systems and sensors are continuously used and exposed to environmental factors, reliable calibration techniques to correct sensor errors in the sensor readings are a must, as drifts in the sensor readings can occur, if sensors are not calibrated regularly [56–58]. Also, a robust calibration procedure preserves data integrity and ensures reliable data insights. Calibration techniques can include comparing the $CO_2$ sensor output with a well-known $CO_2$ reference and applying corrections to align the sensor readings with the $CO_2$ reference. Regular calibration can further maintain accuracy over long periods of system runtime. Therefore, calibration is essential to ensure the long-term reliability and accuracy of IIoT systems for environmental $CO_2$ monitoring and industrial use [56].

### 4.3.6    Data Transmission and Real-time Monitoring

Data transmission from IIoT systems to computing devices like edge or cloud computing servers is another vital function of IIoT systems. Wireless communication technologies like Wi-Fi, LoRaWAN, Zigbee and Bluetooth can be used, depending on the needed data transfer speed [4]. Maintaining $CO_2$ emissions levels, being regulatory compliant, and having a safe working environment, can require real-time data insights. However, choosing the ideal data transmission method requires reliable and secure wireless transmission methods, especially in industrial environments where confidentiality, privacy, and integrity are paramount [4, 12].

### 4.3.7    ML and AI Integration

Integrating AI and ML algorithms into IIoT systems can help analyze big data sets, detect anomalies in data streams, and improve predictive maintenance [59, 60]. Research [61] has found AI models supporting IIoT systems with performance and data accuracy, making such approaches more efficient than previous systems. Furthermore, AI can also provide functions such as expert suggestions based on the IIoT system data and facilitate improved analysis when dealing with extensive data sets [60, 61]. ML algorithms can likewise be integrated to identify potential anomalies in data streams before they become critical [24, 46]. ML algorithms can correspondingly be used to support analyzing historical data and predictive models that can forecast $CO_2$ concentrations based on environmental factors. Furthermore, ML techniques can help to support automated responses to handle $CO_2$ emission changes, such as triggering ventilation by sending real-time alerts to facility managers or support with calibrations by predicting and correcting $CO_2$ sensor drifts dynamically [28, 62].

### 4.4    Use Cases: Intelligent IIoT Systems for Air Quality

IIoT systems can monitor air quality in real-time by measuring air pollutants, temperature, and humidity [62]. Research [63, 64] has demonstrated that IIoT systems are important in industrial environments to detect particle concentrations, and pollutant gases to predict safety level violations, based on ML techniques. The study [65] demonstrates a practical implementation of IIoT devices in vehicles to monitor city air pollution and send the data to cloud servers for analysis. Moreover, the study [7] employed an Arduino-based air quality monitoring system using the versatile MQ-135 gas sensor to detect various gases and $CO_2$ emissions.

## 5    Data Visualization

### 5.1    Visualization Tools and Techniques

Visualization tools are similarly important when considering strategies for gathering data from IIoT systems and CO2 sensors. With the help of visualization tools, the IIoT system data can be visualized into easily understandable formats such as charts, graphs, and heat maps. By applying visualization techniques to the IIoT system output, stakeholders can analyze the collected data, and comprehend real-time changes in CO2 emission levels by improved decision-making [66]. Various visualization tools, such as Tableau, Power BI, and other specialized data-driven dashboards, have been developed to provide user-friendly interfaces for IIoT system outputs [67].

### 5.2    Integration Interfaces and Dashboards

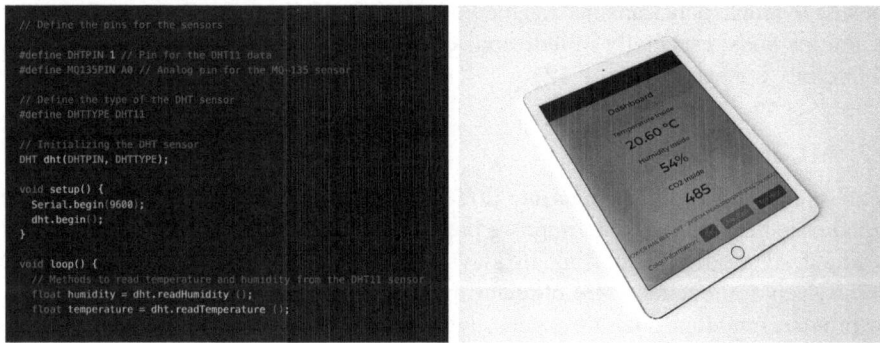

**Fig. 3.** Arduino code and visualization dashboard coded with HTML, CSS, and JavaScript.

The use of web dashboards operated on smartphones, tablets or desktop computers is another modern way of visualizing and showcasing data from IIoT systems [69]. These interfaces, powered by web technologies such as HTML, CSS, and JavaScript, as shown in Fig. 3, enable users and stakeholders to monitor real-time CO2 emissions, and manage the IIoT system data directly within industrial settings. Also, the real-time monitoring capabilities can enhance trust in the accuracy of the IIoT system data, enabling stakeholders to make wireless decisions based on real-time data information from CO2 sensor devices [67–69].

### 5.3    Use Cases: Visualization Techniques in Industrial Environments

Several studies have shown that using IIoT visualization techniques in industrial settings can enhance air quality and emission monitoring, prevent equipment failures, and optimize decision-making processes [70]. One study [71] noted an improvement in tracking equipment utilized by IIoT system technologies. In another study [72], an ML algorithm was designed to show thermal anomalies, giving end-users detailed reports and videos of localized thermal problems. Furthermore, [73] demonstrates that combining IIoT systems with AI and ML techniques combined with human behaviors, can improve IIoT systems visualization output. Moreover, [74] demonstrated AI techniques in IIoT systems to enhance data management by enabling real-time visualization of expert suggestions and assistance.

## 6 Security and Privacy in IIoT Systems

### 6.1 Challenges in IIoT System Security and Privacy

Integrating IIoT systems in industrial environments can present complex security challenges that demand attention [75]. Security challenges include device firmware vulnerabilities, insecure communication protocols, and managing multiple interconnected devices. Also, the diversity of IIoT systems and the lack of standardized security standards further complicate the security aspects of the IIoT system landscape [54]. Physical systems combined with digital controls are especially susceptible to cyberattacks, disrupting operations, and potentially causing significant damage. However, tackling these security challenges can ensure enhanced data integrity and prevent unauthorized data access, especially when dealing with sensitive industrial information [54, 75].

### 6.2 Security Enhancements in IIoT Systems

Despite facing challenges, security measures have made strides in recent years. These advancements involve implementing various protocols and security practices to eliminate security risks associated with IIoT systems, authentication protocols and encryption [54, 76]. Also, integrating blockchain into IIoT systems can create tamper-proof data transactions and maintain data integrity. Moreover, edge computing can support security protocols by processing the IIoT system data locally, and thereby reducing the amount of data transmitted over vulnerable networks [77]. Other recommendations include deploying intrusion detection systems, regularly updating firmware, and adopting security frameworks that address both hardware and software vulnerabilities [78]. Also, to further enhance security practices in IIoT systems, emerging technologies such as ML, AI, and Wireless Sensor Networks (WSN) can support overcoming interoperability, scalability, security, and latency issues in IIoT system architectures [79]. Middleware platforms are also under development to facilitate integration and resolve interoperability issues between IIoT system devices [80, 81]. Conversely, implementing predictive maintenance systems utilized by deep ML algorithms to effectively analyze data from IIoT systems, can also provide support efficiency and system security features [82–84].

### 6.3 Use Cases: Securing IIoT Systems

Security and privacy issues in IIoT systems are critical to industrial operations. Studies have shown that IIoT systems need secure networks to protect data integrity, confidentiality, and availability [85–87]. Also, the connected nature of IIoT systems presents serious security challenges, which testifies strong security protocols as a must [88, 89].

Various security procedures have been proposed to tackle security challenges and comply with IIoT system requirements. These approaches encompass intrusion detection frameworks, authentication protocols based on cryptography and anomaly-based intrusion detection for constrained IIoT networks [90, 91]. Moreover, other studies have also suggested incorporating technologies like blockchain, edge computing, ML, and AI techniques to strengthen the security aspects of IIoT systems [92–94].

# 7 Applications and Use Cases

Despite previous sections that have been looking into IIoT applications and use cases, it is still important to recognize IIoT systems' importance in real-life scenarios. The purpose of this section is to go deeper into how IIoT systems can fit into real-life scenarios and use cases.

## 7.1 Industrial Applications

As mentioned, the Industrial Internet of Things (IIoT) represents an advancement in improving efficiency and productivity across various industrial fields [95, 96]. For instance, factories leverage IIoT systems to oversee equipment status and optimize production, reducing downtime and enhancing effectiveness [71, 97]. Additionally, by employing IIoT systems and connected sensors to monitor manufacturing processes, companies can improve product quality procedures and support employee safety measures effectively [98].

## 7.2 Monitoring the Environment

IIoT systems can monitor the environment by providing real-time data on factors like soil, air, or water conditions. For instance, the research study [99] utilized an IIoT system to track mushroom growing conditions, by integrating sensors to monitor $CO_2$ emissions, humidity, and temperature to ensure optimal growth conditions and improve productivity. In other studies, IIoT systems have been utilized to measure moisture and nutrient levels to facilitate farming techniques and maintain environmental resources effectively [12, 13].

## 7.3 Smart Cities and Buildings

Integrating IIoT systems into cities and buildings can enhance the quality of life and promote efficient resource utilization. IIoT system applications in cities encompass monitoring air quality, managing traffic flow, and optimizing energy usage. Such IIoT systems can further support city planners in implementing strategies to enhance city air quality [100]. Also, integrating seamless air quality regulations based on IIoT systems, can support creating a valuable environment for both workers and residents by adjusting ventilation in real-time, while optimizing energy usage and upholding high air quality standards [101].

# 8   Detailed IIoT System Architecture for CO2 Monitoring

To broaden the understanding of the IIoT system architecture for CO2 monitoring, the following Fig. 4 and Fig. 5 illustrate the features, benefits, and challenges of integrating edge and cloud computing into the IIoT system architecture:

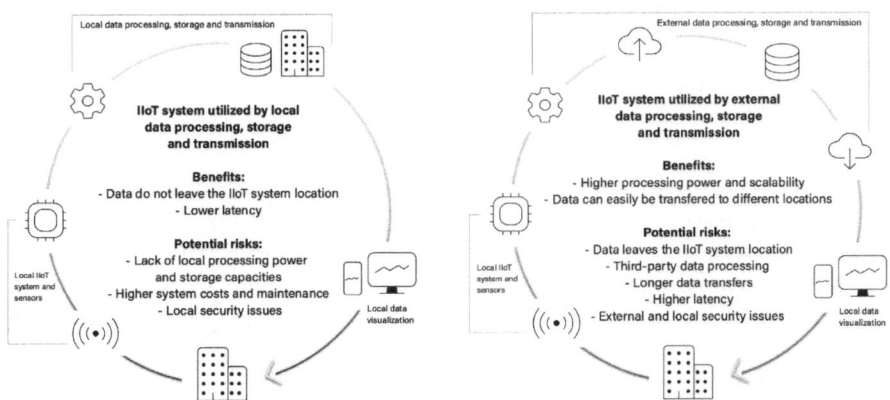

**Fig. 4.**  Detailed IIoT System Architecture for CO2 Monitoring with Edge Computing.

**Fig. 5.**  Detailed IIoT System Architecture for CO2 Monitoring with Cloud Computing.

Figure 4 demonstrates that edge computing is deployed near the IIoT system and CO2 sensor location. This IIoT system architecture allows for local data collection and decision-making, critical for CO2 monitoring scenarios requiring fast response times, such as calibration of sensor drifts, based on real-time CO2 emission levels. However, when utilizing edge computing in IIoT system architectures, there are likewise potential risks, such as lack of processing power and storage capabilities, higher maintenance, system costs, and local security issues. On the other hand, Fig. 5 illustrates cloud computing integrated into the IIoT system architecture. As explained in Fig. 5, cloud computing provides higher processing power, storage options, location independence, and scalability features. However, like with edge computing, it is also important to identify potential risks, which include data leaving the IIoT system location, third-party data handling, longer data transfers, higher latency, and security risks both locally and externally.

As Fig. 4 and Fig. 5 demonstrate, the two IIoT system architectures offer different benefits and challenges. However, even with the figures' dual approach, IIoT system architectures can also be presented in a hybrid way, based on a blend of potential benefits and system risks.

# 9   System Prototype for Monitoring CO2 Emission Levels

Building on the previous understanding, the following section presents an environmental monitoring system developed and tested as a prototype experiment. The aim was to explore the feasibility of using IIoT system devices to monitor CO2 emission levels, focusing on real-time data visualization and a low-cost system design. The development process involved integrating an Arduino-based microcontroller (ESP8266) with connected sensors, encased in a 3D-printed enclosure. However, while the MQ-135 sensor utilized in this prototype experiment is widely available and cost-effective for conducting $CO_2$ emissions, it is essential to note that this specific sensor has limitations in accuracy, especially when compared to more advanced $CO_2$ sensors. The system was moreover tested for accuracy and reliability according to data processing and acquisition. The steps undertaken in the development process can be detailed as follows:

## 9.1 Hardware Integration

The primary hardware components used in the system included:

- **ESP8266 Microcontroller:** An Arduino Wi-Fi-based microcontroller, functioning as the core processing and data logging unit.
- **MQ-135 sensor:** An analog gas sensor used to measure $CO_2$ emission levels.
- **BME680 sensor:** A multifunction sensor to track temperature and humidity levels simultaneously.

## 9.2 Software Integration

The system software development involved the following key areas:

- **Data Logging:** Data logging mechanisms were implemented based on edge computing techniques, utilized by the ESP8266's SPIFFS (SPI Flash Files System) to locally store and log the collected emission data on the ESP8266's file system, as well as store the WIFI credentials for the system.
- **Time Synchronization:** Time synchronization was handled using NTP (Network Time Protocol) to ensure that the data was timestamped accurately in the ESP8266's EEPROM, and for adjusting offsets based on system runtime.
- **Data Processing and Calibration:** An initial sensor calibration was implemented into the system code, based on an outside $CO_2$ reference of around 400ppm. Additionally, a series of calibration offsets was implemented to dynamically maintain sensor calibrations, depending on different stages of system runtime.
- **Webserver and User Interface:** An onboard webserver was deployed directly onto the ESP8266 microcontroller, allowing for hosting a visualization dashboard and user interface with real-time data updates.
- **Visualization Dashboard:** A visualization dashboard and user interface were developed utilizing a combination of HTML, CSS, and JavaScript for cross-platform visualization features, including functionalities to download system data, stored in the ESP8266's file system.

## 9.3 Test and System Validation

The system underwent a series of tests to understand system performance, which included:

- **Stability Testing:** Continuous monitoring over an extended period of runtime to test sensor reading and system stability.

- **Accuracy Validation:** Comparing the sensor outputs with other available instruments to understand the accuracy of the system measurements.
- **Environmental Testing:** Placing the system under different environmental conditions to test the robustness and reliability of system measurements and visualizations.

**Fig. 6.** Parts of the System Prototype for Monitoring CO2 Emission Levels.

## 9.4 Final System Remarks

The development process of the system prototype, as seen in Fig. 6, demonstrates the feasibility of creating a low-cost environmental monitoring system utilizing IIoT components, combined with visualization dashboards, suitable for both research and practical applications in various domains. However, as the MQ-135 sensor is not an untainted CO2 sensor, and therefore can be affected by other gases, it is assumed that systems based solely on the MQ-135 sensor for monitoring CO2 emissions, may not deliver truthful measurements over long periods of runtime. It is therefore recommended to compare system designs with other and more advanced CO2 sensors, to get a more reliable picture of the system measurements. Despite these limitations, it is noteworthy that the prototype experiment demonstrates the feasibility of a low-cost system design for CO2 monitoring and real-time visualization, highlighting the potential for future development with more precise CO2 sensors.

## 10  Challenges and Future Directions

Despite advancements in IIoT systems, future research should focus on functions that can improve the effectiveness and the adoption of IIoT systems for monitoring CO2 levels in industrial environments, which include:

### 10.1  Limitations of IIoT Systems

Implementing IIoT systems can be expensive as it involves investing in hardware, software, and ongoing maintenance to ensure reliable operation [102]. For instance, integrating IIoT systems into a manufacturing facility may require high initial investment, based on factors like lack of internal expertise [9]. Also, resistance within the workforce to adopting into the IIoT ecosystem can hinder implementation. Other various factors such as limitations in IT infrastructure, concerns about data security, and uncertainty about job displacement, may also act as barriers to the integration of IIoT system solutions [103, 104]. Additionally, industries operating in areas with limited internet connectivity may also face issues with transmission delays and system downtime that can impact IIoT systems' efficiency, due to their reliance on data connections [105, 106].

### 10.2  Challenges and Improvements in IIoT Systems

Implementing IIoT systems poses several challenges, including sensor compatibility, data security and system scalability. One major obstacle is ensuring collaboration among all IIoT system devices used in system setups for monitoring CO2 emissions levels. Device manufacturers' implementation of different system protocols can also complicate integration efforts and prevent seamless data sharing [54, 102, 107]. As a result, it is essential to explore strong communication protocols and encryption techniques to safeguard industrial data against external cyber threats. Moreover, handling the vast amount of data generated by IIoT systems also demands storage capabilities and effective data management to ensure efficient data flow. Furthermore, effectively utilizing real-time data insights without system delays is another challenge [23]. Focusing on data protocols that facilitate seamless integration and communication among IIoT systems and devices on a larger scale is equally needed. In addition, prioritizing energy sources for power delivery to IIoT systems is likewise crucial [9]. Further research should also be focused on developing enhanced AI suggestion services or ML algorithms to streamline large data and analysis processes [61]. Additionally, future research should focus on blockchain technology to improve IIoT system security through transparent data transactions [85]. Although even if blockchain technology seems promising, its integration into IIoT systems may also present challenges, such as system complexity and further delays in the data streams. As a result, future research should concentrate on improving IIoT system protocols to reduce system delays while maintaining high-security standards [54].

#### 10.2.1  Scalability and Integration

Scalability and integration are critical elements when dealing with IIoT system integrations. While IIoT systems work quite effectively in controlled settings, the possibility of expanding and integrating IIoT systems into more challenging environments, like urban areas or complex industrial facilities, can present further challenges to the system landscape [9, 107]. Future research should focus on developing IIoT system architectures that can process large volumes of data and integrate seamlessly with existing infrastructure within industries [9, 108]. Researchers and IIoT system architects should also focus on developing enhanced IIoT systems, based on strategies to ensure performance and reliability at large system scales and as IIoT networks expand [44].

### 10.2.2 User-Friendly Interfaces and Employee Training

User-friendly interfaces and training employees are also key for future IIoT systems to succeed, by prioritizing the development of easy-to-understand visualization interfaces and making solutions more accessible for non-technical users [13, 66]. In addition, the successful integration of IIoT systems further heavily depends on employee readiness and the ability to use IIoT systems effectively [103]. Future research should focus on identifying strategies for user training and intuitive user interfaces that promote the intuitive adoption of IIoT system integrations [9, 104].

### 10.2.3 Real-Time Decision Making

Stakeholders' ability to make immediate real-time decisions based on the IIoT system data and to act when anomalies are detected should also be optimized [12]. Future research should focus on defining data processing and visualization techniques utilized by AI systems to provide expert commentary and suggestions on data analysis and enable ML algorithms to dynamically calibrate real-time fluctuations in $CO_2$ sensors to deliver accurate real-time system information [56].

### 10.2.4 Economic Feasibility

Cost can also be a barrier to implementing IIoT systems for monitoring $CO_2$ emission levels, as it may be necessary for businesses to seek assistance in investing and embracing IIoT system technologies for $CO_2$ monitoring [9, 104]. Future studies could concentrate on cutting down the cost of IIoT systems and sensor devices by utilizing budget sensor technologies or software systems based on open-source platforms [27]. Future research demonstrating cost-effectiveness and return on investment could involve economic and cost-benefit analyses to illustrate how IIoT systems for $CO_2$ monitoring can result in long-term savings and sustainable advantages, as well as promote wider adoption within organizations and industries [9].

### 10.2.5 Long-Term Data Management

Dealing with the vast amount of data produced by IIoT systems and connected sensors is also challenging [9]. Managing data from multiple IIoT systems and sensors involves working with multiple data streams from various sources, which can require improved data storage solutions to cope with the vast amount of data that industrial environments generate [107, 108]. Additionally, enhancing data management systems and predictive maintenance strategies based on IIoT system data may require investigating solutions with faster data transfer speeds, along with improved cloud computing and data management options [109].

### 10.2.6   Practical and Comparative Analysis

Lastly, a comparative analysis of IIoT systems and connected CO2 sensors in real-world and industrial settings might also be an additional aspect to consider. A comparative analysis would facilitate the evaluation and effectiveness of different IIoT system architectures and connected CO2 sensors - and especially where CO2 levels are highly varied. Future research should focus on experimenting with different IIoT systems architectures directly in industrial environments, to gain an added understanding of performance matrices, such as latency, data accuracy, and system reliability, to determine the paramount balance between IIoT systems and CO2 sensor devices.

## 11   Conclusion

This paper has examined how IIoT systems can monitor, measure, and visualize CO2 emissions levels in industries and the environment. IIoT systems can monitor CO2 levels and air quality in real-time by analyzing CO2 emission data collected, processed, and visualized by IIoT systems and sensors [2, 4, 16]. Using advanced CO2 sensors, data fusion, AI, and ML techniques, CO2 monitoring can increase CO2 emission accuracy and efficiency in industries [1, 59, 60]. Moreover, utilizing IIoT systems to measure CO2 emissions extends beyond industrial settings and can further be utilized in agriculture, urban environments, and buildings where CO2 emission levels are vital and benefit future industries and society as a whole [4, 8, 12].

### 11.1   Future Perspectives

Looking ahead there are plenty of ways in which IIoT systems can benefit from monitoring CO2 emission levels. However, future research should focus on more energy-efficient and scalable IIoT systems, data security, and advanced technologies, such as edge computing and blockchain, to monitor CO2 emission processes [9, 23, 86]. AI-assisted services and ML algorithms to support predicting and detecting anomalies and system maintenance should also be prioritized [24, 46]. Likewise, focusing on standardized communication protocols and interoperability frameworks, which can facilitate the connection of devices from diverse system vendors, should also be emphasized, allowing IIoT systems to be more adoptable in supporting CO2 emission control across industries [9, 54].

## References

1. Le, T., Oktian, Y., Kim, H.: XGBoost for imbalanced multiclass classification-based industrial internet of things intrusion detection systems. Sustainability **14**(14), 8707 (2022). https://doi.org/10.3390/su14148707

2. Mofidul, R.B., Rahman, M.H., Jang, Y.M.: Real-time energy data acquisition, anomaly detection, and monitoring system: implementation of a secured, robust, and integrated global IIoT infrastructure with edge and cloud AI. Sensors **22**(22), 8980–8980 (2022). https://doi.org/10.3390/s22228980
3. Kaur, M., Khan, M., Gupta, S., Alsaeedi, A.: Adoption of blockchain with 5G networks for industrial IoT: recent advances, challenges, and potential solutions. IEEE Access **10**, 981–997 (2022). https://doi.org/10.1109/access.2021.3138754
4. Mateev, V., Marinova, I.: Distributed Internet of Things System for $CO_2$ Monitoring with LoRaWAN (2021). https://doi.org/10.1109/ELECTRONICA52725.2021.9513682
5. Deepan, S., Buradkar, M., Akhila, P., Kumar, K.S., Sharma, M.K. and Chakravarthi, M.K.: AI-powered predictive maintenance for industrial IoT systems. In: Proceedings of the 2024 (2024). https://doi.org/10.1109/ACCAI61061.2024.10601983
6. Bansod, N.V., Hore, U.W.: IoT based air quality monitoring system, pp. 1106–1114 (2021). https://doi.org/10.48175/IJARSCT-1536
7. Abbas, F.N., Saadoon, M.M., Abdalrdha, Z.K., Abud, E.N.: Capable of gas sensor MQ-135 to monitor the air quality with Arduino Uno. Int. J. Eng. Res. Technol. (2020). https://doi.org/10.37624/IJERT/13.10.2020.2955-2959
8. Sisinni, E., Saifullah, A., Han, S., Jennehag, U., Gidlund, M.: Industrial internet of things: challenges, opportunities, and directions. IEEE Trans. Ind. Inf. **14**(11), 4724–4734 (2018). https://doi.org/10.1109/tii.2018.2852491
9. Wójcicki, K., Biegańska, M., Paliwoda, B., Górna, J.: Internet of things in industry: research profiling, application, challenges and opportunities—A review. Energies **15**(5), 1806 (2022). https://doi.org/10.3390/en15051806
10. Saragea, Ş, Toma, M., Micu, D., Fratila, G., Badea, G.: Software strategy for internal combustion engine and electric motor control on a hybrid electric vehicle equipped with belt starter generator and automated manual transmission. IOP Conf. Ser. Mater. Sci. Eng. **1235**(1), 012035 (2022). https://doi.org/10.1088/1757-899x/1235/1/012035
11. Salhaoui, M., Guerrero-González, A., Arioua, M., Ortiz, F., Oualkadi, A., Torregrosa, C.: Smart industrial IoT monitoring and control system based on UAV and cloud computing applied to a concrete plant. Sensors **19**(15), 3316 (2019). https://doi.org/10.3390/s19153316
12. Ullo, S., Sinha, G.: Advances in smart environment monitoring systems using IoT and sensors. Sensors **20**(11), 3113 (2020). https://doi.org/10.3390/s20113113
13. Maulana, G.G., Aminah, S., Berlliyanto, A.N.: Implementation of a production monitoring system using IIoT based on mobile application. Jurnal RESTI (Rekayasa Sistem dan Teknologi Informasi) 7(5) (2023). https://doi.org/10.29207/resti.v7i5.5221
14. Al-Rubaye, S., Kadhum, E., Ni, Q., Anpalagan, A.: Industrial internet of things driven by SDN platform for smart grid resiliency. IEEE Internet Things J. **6**(1), 267–277 (2019). https://doi.org/10.1109/jiot.2017.2734903
15. Jameel, F., Javaid, U., Khan, W., Aman, M., Pervaiz, H., Jäntti, R.: Reinforcement learning in blockchain-enabled IIoT networks: a survey of recent advances and open challenges. Sustainability **12**(12), 5161 (2020). https://doi.org/10.3390/su12125161
16. Sarkar, A., Ghosh, D., Ganguly, K., Ghosh, S., Saha, S.: Exploring IoT for real-time CO2 monitoring and analysis. arXiv.org, vol. abs/2308.03780 (2023).https://doi.org/10.48550/arxiv.2308.03780
17. Shi, W., Cao, J., Zhang, Q., Li, Y., Xu, L.: Edge computing: vision and challenges. IEEE Internet Things J. **3**(5), 637–646 (2016). https://doi.org/10.1109/JIOT.2016.2579198
18. Mendes, L., Ogink, N., Edouard, N., Dooren, H., Tinôco, I., Mosquera, J.: NDIR gas sensor for spatial monitoring of carbon dioxide concentrations in naturally ventilated livestock buildings. Sensors **15**(5), 11239–11257 (2015). https://doi.org/10.3390/s150511239

19. Macagga, R., Asante, M., Sossa, G., Antonijevic, D., Dubbert, M., Hoffmann, M.: Validation and field application of a low-cost device to measure CO2 and ET fluxes. Environ. Sci. Technol. **57**(5), 2093–2101 (2023). https://doi.org/10.5194/amt-17-1317-2024

20. Pino, H., Pastor, V., Grimalt-Álvaro, C., López, V.: Measuring CO2 with an Arduino: creating a low-cost, pocket-sized device with flexible applications that yield benefits for students and schools. J. Chem. Educ. **96**(2), 377–381 (2018). https://doi.org/10.1021/acs.jchemed.8b00473

21. Gunawan, T., Munir, Y., Kartiwi, M., Mansor, H.: Design and implementation of a portable outdoor air quality measurement system using Arduino. Int. J. Electr. Comput. Eng. **8**(1), 280–288 (2018). https://doi.org/10.11591/ijece.v8i1.pp280-290

22. Kelechi, A., et al.: Design of a low-cost air quality monitoring system using Arduino and ThingSpeak. Comput. Mater. Continua **70**(1), 151–169 (2022). https://doi.org/10.32604/cmc.2022.019431

23. Qiu, T., Chi, J., Zhou, X., Ning, Z., Atiquzzaman, M., Wu, D.: Edge computing in industrial internet of things: architecture, advances and challenges. IEEE Commun. Surv. Tutor. **22**(4), 2462–2488 (2020). https://doi.org/10.1109/comst.2020.3009103

24. Wu, D., Jiang, Z., Xie, X., Wei, X., Yu, W., Li, R.: LSTM learning with Bayesian and Gaussian processing for anomaly detection in industrial IoT. IEEE Trans. Ind. Inform. **16**(8), 5244–5253 (2020). https://doi.org/10.1109/tii.2019.2952917

25. Struzik, M., et al.: A simple and fast electrochemical CO2 sensor based on Li7La3Zr2O12 for environmental monitoring. Adv. Mater. (2018). https://doi.org/10.1002/ADMA.201804098

26. Chen, H., Markham, J.: Using microcontrollers and sensors to build an inexpensive Co2 control system for growth chambers. Appl. Plant Sci. **8**(10) (2020). https://doi.org/10.1002/aps3.11393

27. Jha, R.K.: Non-dispersive infrared gas sensing technology: a review. IEEE Sens. J. **22**(1), 6–15 (2022). https://doi.org/10.1109/JSEN.2021.3130034

28. Yong, Y., Zhang, C., He, C., Wang, X., Huang, J., Deng, J.: A review on applications of capacitive displacement sensing for capacitive proximity sensor. IEEE Access **8**, 45325–45342 (2020). https://doi.org/10.1109/access.2020.2977716

29. Gu, M., et al.: Portable TDLAS sensor for online monitoring of CO2 and H2O using a miniaturized multi-pass cell. Sensors **23**(4), 2072 (2023). https://doi.org/10.3390/s23042072

30. Ji, S., Lee, J.H., Lee, S.-H., Hong, S.: Solid-state amperometric CO2 sensor using a sodium ion conductor. J. Eur. Ceram. Soc. **24**(6), 1431–1434 (2004). https://doi.org/10.1016/s0955-2219(03)00430-8

31. Graef, E.W., Munje, R.D., Prasad, S.: A robust electrochemical CO2 sensor utilizing room temperature ionic liquids. IEEE Trans. Nanotechnol. **16**(5), 826–831 (2017). https://doi.org/10.1109/TNANO.2017.2672599

32. Rivadeneyra, A., López-Villanueva, J.: Recent advances in printed capacitive sensors. Micromachines **11**(4), 367 (2020). https://doi.org/10.3390/mi11040367

33. Endres, H.-E., et al.: A capacitive CO2 sensor system with suppression of the humidity interference. Sens. Actuators B Chem. (1999). https://doi.org/10.1016/S0925-4005(99)00060-X

34. Palzer, S.: Photoacoustic-based gas sensing: a review. Sensors **20**(9), 2745 (2020). https://doi.org/10.3390/s20092745

35. Hannon, A., Li, J.: Solid state electronic sensors for detection of carbon dioxide. Sensors (2019). https://doi.org/10.3390/s19183848

36. Kostinek, J., et al.: Adaptation and performance assessment of a quantum and interband cascade laser spectrometer for simultaneous airborne in situ observation of $CH_4$, $C_2H_6$, $CO_2$, CO and $N_2O$. Atmos. Meas. Tech. **12**, 1767–1779 (2019). https://doi.org/10.5194/amt-12-1767-2019

37. Liu, Z., et al.: Midinfrared sensor system based on tunable laser absorption spectroscopy for dissolved carbon dioxide analysis in the South China Sea: system-level integration and deployment (2020). https://doi.org/10.1021/acs.analchem.0c00327

38. Sully, M., et al.: All-fiber $CO_2$ sensor using hollow core PCF operating in the 2 $\mu$m region (2018). https://doi.org/10.3390/S18124393

39. Zhou, Z., Xu, Y., Qiao, C., Liu, L., Jia, Y.: A novel low-cost gas sensor for $CO_2$ detection using polymer-coated fiber Bragg grating. Sens. Actuators B Chem. (2021). https://doi.org/10.1016/j.snb.2021.129482

40. Huang, X., Shen, Q., Liu, J., Yang, N., Zhao, G.: A $CO_2$ adsorption-enhanced semiconductor/metal-complex hybrid photo electrocatalytic interface for efficient format production. Energy Environ. Sci. 9(10), 3161–3171 (2016). https://doi.org/10.1039/C6EE00968A

41. Maeda, K.: Metal-complex/semiconductor hybrid photocatalysts and photoelectrodes for $CO_2$ reduction driven by visible light. Adv. Mater. (2019). https://doi.org/10.1002/ADMA.201808205

42. Han, T.-R., Joshi, V., Gabriel, J.C., Grüner, G.: Nanoelectronic carbon dioxide sensors. Adv. Mater. 16(22), 2049–2052 (2004). https://doi.org/10.1002/adma.200400322

43. Samotaev, N., et al.: Thermal conductivity gas sensors for high-temperature applications. Micromachines 15(1), 138 (2024). https://doi.org/10.3390/mi15010138

44. Lim, J.: Scalable fog computing orchestration for reliable cloud task scheduling. Appl. Sci. 11(22), 10996 (2021). https://doi.org/10.3390/app112210996

45. Chen, X., Jiao, L., Li, W., Fu, X.: Efficient multi-user computation offloading for mobile-edge cloud computing. IEEE ACM Trans. Netw. 24(5), 2795–2808 (2016). https://doi.org/10.1109/TNET.2015.2487344

46. Sui, Q., Liu, X.: Edge computing and AIoT-based network intrusion detection mechanism. Internet Technol. Lett. 6(5) (2021). https://doi.org/10.1002/itl2.324

47. Johirul, I., Tanesh, K., Ivana, K., Erkki, H.: Resource-aware dynamic service deployment for local IoT edge computing: healthcare use case. IEEE Access (2021). https://doi.org/10.1109/ACCESS.2021.3102867

48. Wang, N., Varghese, B., Matthaiou, M., Nikolopoulos, D.: ENORM: a framework for edge node resource management. IEEE Trans. Serv. Comput. (2019). https://doi.org/10.1109/TSC.2017.2753775

49. Liu, S.: Satellite-air-terrestrial cloud edge collaborative networks: architecture, multi-node task processing and computation. Intell. Autom. Soft Comput. 37(3), 2651–2668 (2023). https://doi.org/10.32604/iasc.2023.038477

50. Samanta, A., Ha, T., Nguyen, T.: Distributed resource distribution and offloading for resource-agnostic microservices in industrial IoT. IEEE Trans. Veh. Technol. 72(1), 1184–1195 (2023). https://doi.org/10.1109/tvt.2022.3206137

51. Han, P., Wang, S., Leung, K.: Capacity analysis of distributed computing systems with multiple resource types. In: IEEE Wireless Communications and Networking Conference (2020). https://doi.org/10.1109/WCNC45663.2020.9120531

52. Tank, B., Gandhi, V.: A comparative study on cloud computing, edge computing, and fog computing. Adv. Transdiscipl. Eng. (2023). https://doi.org/10.3233/atde221329

53. Peng, K.H., Huang, H., Zhao, B., Jolfaei, A., Xu, X., Bilal, M.: Intelligent computation offloading and resource allocation in IIoT with end-edge-cloud computing using NSGA-III. IEEE Trans. Netw. Sci. Eng. (2022). https://doi.org/10.1109/tnse.2022.3155490

54. Gebremichael, T., et al.: Security and privacy in the industrial internet of things: current standards and future challenges. IEEE Access 8, 152351–152366 (2020). https://doi.org/10.1109/access.2020.3016937

55. Yuehua, L., Tharam, S.D., Wenjin, Y., Wenny, R., Fahed, M.: Noise removal in the presence of significant anomalies for industrial IoT sensor data in manufacturing. IEEE Internet Things J. (2020). https://doi.org/10.1109/JIOT.2020.298147Ω

56. Kim, J., Shusterman, A.A., Lieschke, K.J., Newman, C., Cohen, R.C.: The BErkeley Atmospheric $CO_2$ Observation Network: field calibration and evaluation of low-cost air quality sensors. Atmos. Meas. Tech. **11**(4), 1937–1946 (2017). https://doi.org/10.5194/AMT-11-1937-2018

57. Ferrer-Cid, P., Barceló-Ordinas, J., García-Vidal, J., Ripoll, A., Viana, M.: Multisensor data fusion calibration in IoT air pollution platforms. IEEE Internet Things J. **7**(4), 3124–3132 (2020). https://doi.org/10.1109/jiot.2020.2965283

58. Liu, X., Jia, M., Zhou, M., Wang, B., Durrani, T.: Integrated cooperative spectrum sensing and access control for cognitive industrial internet of things. IEEE Internet Things J. **10**(3), 1887–1896 (2023). https://doi.org/10.1109/jiot.2021.3137408

59. Venkatasubramanian, S., Raja, S., Dwivedi, J., Sathiaparkavi, J., Modak, S., Kejela, M.: Fault diagnosis using data fusion with ensemble deep learning technique in IIoT. Math. Probl. Eng. **2022**, 1–8 (2022). https://doi.org/10.1155/2022/1682874

60. Nozari, H., Szmelter-Jarosz, A., Ghahremani-Nahr, J.: Analysis of the challenges of artificial intelligence of things (AIoT) for the smart supply chain (case study: FMCG industries). Sensors **22**(8), 2931 (2022). https://doi.org/10.3390/s22082931

61. Adi, E., Anwar, A., Baig, Z., Zeadally, S.: Machine learning and data analytics for the IoT. Neural Comput. Appl. **32**(20), 16205–16233 (2020). https://doi.org/10.1007/s00521-020-04874-y

62. Ng, W., Dahari, Z.: Enhancement of real-time IoT-based air quality monitoring system. Int. J. Power Electron. Drive Syst. (IJPEDS) **11**(1), 390 (2020). https://doi.org/10.11591/ijpeds.v11.i1.pp390-397

63. Evison, F.: IoT enabled environmental toxicology for air pollution monitoring using AI techniques. Environ. Res. (2022). https://doi.org/10.1016/j.envres.2021.112574

64. Pramanik, J., Samal, A.K., Pani, S.K., Chakraborty, C.: Elementary framework for an IoT based diverse ambient air quality monitoring system. Multimed. Tools Appl. **81** (2021). https://doi.org/10.1007/s11042-021-11285-1

65. Alvear-Puertas, V., Burbano-Prado, Y.A., Rosero-Montalvo, P.D., Tözün, P., Marcillo, F., Hernandez, W.: Smart and portable air-quality monitoring IoT low-cost devices in Ibarra city, Ecuador. Sensors (2022). https://doi.org/10.3390/s22187015

66. Zhang, Q.: The impact of interactive data visualization on decision-making in business intelligence. In: Advances in Economics, Management and Political Sciences (2024). https://doi.org/10.54254/2754-1169/87/20241056

67. Jeong, Y., Joo, H., Hong, G., Shin, D., Lee, S.: AVIoT: web-based interactive authoring and visualization of indoor internet of things. IEEE Trans. Consum. Electron. (2015). https://doi.org/10.1109/TCE.2015.7298088

68. Babovic, Z., Protic, J., Milutinovic, V.: Web performance evaluation for internet of things applications. IEEE Access (2016). https://doi.org/10.1109/ACCESS.2016.2615181

69. Lomotey, R.K., Pry, J., Chai, C.: Traceability and visual analytics for the Internet-of-Things (IoT) architecture. World Wide Web (2018). https://doi.org/10.1007/s11280-017-0461-1

70. Ungurean, I., Gaitan, N.: A software architecture for the industrial internet of things—A conceptual model. Sensors **20**(19), 5603 (2020). https://doi.org/10.3390/s20195603

71. Невлюдов, И., Yevsieiev, V., Maksymova, S., Demska, N., Starodubcev, N., Klymenko, O.: Monitoring system development for equipment upgrade for IIoT (2023). https://doi.org/10.1109/mees61502.2023.10402532

72. Ghazal, M., Basmaji, T., Yaghi, M., Alkhedher, M., Mahmoud, M., El-Baz, A.S.: Cloud-based monitoring of thermal anomalies in industrial environments using AI and the internet of robotic things. Sensors (2020). https://doi.org/10.3390/S20216348

73. Ullah, F.U.M., et al.: AI-assisted edge vision for violence detection in iot-based industrial surveillance networks. IEEE Trans. Ind. Inform. **18**(8), 5359–5370 (2022). https://doi.org/10.1109/TII.2021.3116377
74. Lavalle, A., Teruel, M.A., Maté, A., Trujillo, J.: Fostering sustainability through visualization techniques for real-time IoT data: a case study based on gas turbines for electricity production. Sensors (2020). https://doi.org/10.3390/S20164556
75. Shitharth, S., et al.: An artificial intelligence lightweight blockchain security model for security and privacy in IIoT systems. J. Cloud Comput. **12**(1) (2023). https://doi.org/10.1186/s13677-023-00412-y
76. Lara, E., Aguilar, L., Sánchez, M., García, J.: Lightweight authentication protocol for M2M communications of resource-constrained devices in industrial internet of things. Sensors **20**(2), 501 (2020). https://doi.org/10.3390/s20020501
77. Wu, Y., Dai, H., Wang, H.: Convergence of blockchain and edge computing for secure and scalable IIoT critical infrastructures in industry 4.0. IEEE Internet Things J. **8**(4), 2300–2317 (2021). https://doi.org/10.1109/JIOT.2020.3025916
78. Fun, T., Samsudin, A.: Recent technologies, security countermeasure and ongoing challenges of industrial internet of things (IIoT): a survey. Sensors **21**(19), 6647 (2021). https://doi.org/10.3390/s21196647
79. Mirani, A., Velasco-Hernandez, G., Awasthi, A., Walsh, J.: Key challenges and emerging technologies in industrial IoT architectures: a review. Sensors **22**(15), 5836 (2022). https://doi.org/10.3390/s22155836
80. John, J., Ghosal, A., Margaria, T., Pesch, D.: DSLs and middleware platforms in a model-driven development approach for secure predictive maintenance systems in smart factories. In: Margaria, T., Steffen, B. (eds.) ISoLA 2021. LNTCS, vol. 13036, 146–161. Springer, Cham (2021). https://doi.org/10.1007/978-3-030-89159-6_10
81. Paniagua, C., Delsing, J.: Industrial frameworks for internet of things: a survey. IEEE Syst. J. **15**(1), 1149–1159 (2021). https://doi.org/10.1109/JSYST.2020.2993323
82. De, S., Bermúdez-Edo, M., Xu, H., Cai, Z.: Deep generative models in the industrial internet of things: a survey. IEEE Trans. Ind. Inform. **18**(9), 5728–5737 (2022). https://doi.org/10.1109/TII.2022.3155656
83. Pal, S., Jadidi, Z.: Analysis of security issues and countermeasures for the industrial internet of things. Appl. Sci. **11**(20), 9393 (2021). https://doi.org/10.3390/app11209393
84. Sree, T.: Role of fog-assisted industrial internet of things: a systematic review. Trans. Emerg. Telecommun. Technol. **33**(12) (2022). https://doi.org/10.1002/ett.4611
85. Singh, A., Raghav, S., Kumar, P., Ashish, S.: Blockchain and edge computing for industrial internet of things (IIoT) applications. Asian J. Converg. Technol. **8**(1), 107 (2022). https://doi.org/10.33130/ajct.2022v08i01.016
86. Huang, J., Kong, L., Chen, G., Wu, M.-Y., Li, X., Zeng, P.: Towards secure industrial IoT: blockchain system with credit-based consensus mechanism. IEEE Trans. Ind. Inform. (2019). https://doi.org/10.1109/tii.2019.2903342
87. Abosata, N., Al-Rubaye, S., Tsourdos, A., Emmanouilidis, C.: Internet of things for system integrity: a comprehensive survey on security, attacks and countermeasures for industrial applications. Sensors (2021). https://doi.org/10.3390/s21113654
88. Tange, K., Donno, M.D., Fafoutis, X., Dragoni, N.: A systematic survey of industrial internet of things security: requirements and fog computing opportunities. IEEE Commun. Surv. Tutor. (2020). https://doi.org/10.1109/comst.2020.3011208
89. Alshahrani, H., Khan, A., Rizwan, M., Al Reshan, M.S., Sulaiman, A., Shaikh, A.: Intrusion detection framework for industrial internet of things using software defined network. Sustainability (2023). https://doi.org/10.3390/su15119001

90. Zhao, X., Li, D., Li, H.: Practical three-factor authentication protocol based on elliptic curve cryptography for industrial internet of things. Sensors (2022). https://doi.org/10.3390/s22 197510

91. Essop, I., Ribeiro, J.C., Παπαϊωάννου, M., Zachos, G., Μαντάς, Γ., Rodríguez, J.: Generating datasets for anomaly-based intrusion detection systems in IoT and industrial IoT networks. Sensors (2021). https://doi.org/10.3390/s21041528

92. Aslani, M., Amin, O., Nawab, F., Shihada, B.: Rethinking blockchain integration with the industrial internet of things. IEEE Internet Things Mag. (2020). https://doi.org/10.1109/iotm.0001.1900079

93. Jedidi, A.: Dynamic trust security approach for edge computing-based mobile IoT devices using artificial intelligence. Eng. Res. Express (2024). https://doi.org/10.1088/2631-8695/ad43b5

94. Sun, W.-L., Tang, Y., Huang, Y.-L.: HiRAM: a hierarchical risk assessment model and its implementation for an industrial internet of things in the cloud. Softw. Test. Verif. Reliab. (2023). https://doi.org/10.1002/stvr.1847

95. Mantravadi, S., Schnyder, R., Møller, C., Brunoe, T.D.: Securing IT/OT links for low power IIoT devices: design considerations for industry 4.0. IEEE Access (2020). https://doi.org/10.1109/access.2020.3035963

96. Tange, K., Donno, M., Fafoutis, X., Dragoni, N.: A systematic survey of industrial internet of things security: requirements and fog computing opportunities. IEEE Commun. Surv. Tutor. **22**(4), 2489–2520 (2020). https://doi.org/10.1109/comst.2020.3011208

97. Misra, S., Roy, C., Sauter, T., Mukherjee, A., Maiti, J.: Industrial internet of things for safety management applications: a survey. IEEE Access **10**, 83415–83439 (2022). https://doi.org/10.1109/access.2022.3194166

98. Annusharma, M.A., Maharani, C.: Enhanced factory safety system using IoT. Int. J. Adv. Res. Sci. Commun. Technol. 1–4 (2024). https://doi.org/10.48175/ijarsct-19001

99. Chong, J., Chew, K., Peter, A., Ting, H., Show, P.: Internet of things (IoT)-based environmental monitoring and control system for home-based mushroom cultivation. Biosensors **13**(1), 98 (2023). https://doi.org/10.3390/bios13010098

100. Zhou, C., Li, S., Wang, S.: Examining the impacts of urban form on air pollution in developing countries: a case study of China's megacities. Int. J. Environ. Res. Public Health **15**(8), 1565 (2018). https://doi.org/10.3390/ijerph15081565

101. Suryawanshi, R., Garje, R., Ghodake, S., Nadar, H., Ingle, P., Shaikh, I.: IoT based real-time environment monitoring and safety for factory workplace. In: Proceedings of the 2024, pp. 1-6 (2024). https://doi.org/10.1109/ic-cgu58078.2024.10530721

102. He, W., Da Xu, L.: Integration of distributed enterprise applications: a survey. IEEE Trans. Ind. Inf. **10**(1), 35–42 (2014). https://doi.org/10.1109/TII.2012.2189221

103. Gadekar, R., Sarkar, B., Gadekar, A.: Model development for assessing inhibitors impacting industry 4.0 implementation in Indian manufacturing industries: an integrated ISM-fuzzy MICMAC approach. Int. J. Syst. Assur. Eng. Manag. **15**(2), 646–671 (2022). https://doi.org/10.1007/s13198-022-01691-5

104. Zhuankhan, A., Renken, J.: Exploring the determinants of successful IoT adoption: the case of German manufacturing. Int. J. Innov. Technol. Manag. **20**(07), 2350046 (2023). https://doi.org/10.1142/S0219877023500463

105. Reddy, G., Rangaswamy, K., Sudhakara, M., Anjaiah, P., Madhavi, K.: Towards the protection and security in fog computing for industrial internet of things. In: Advances in Computer and Electrical Engineering, pp. 17–32 (2021). https://doi.org/10.4018/978-1-7998-3375-8.ch002

106. Harjula, E., Artemenko, A., Forsström, S.: Edge computing for industrial IoT: challenges and solutions. In: Mahmood, N.H., Marchenko, N., Gidlund, M., Popovski, P. (eds.) Wireless

Networks and Industrial IoT, pp. 225–240. Springer, Cham (2021). https://doi.org/10.1007/978-3-030-51473-0_12

107. Lee, E., Seo, Y.D., Oh, S.R., Kim, Y.G.: A survey on standards for interoperability and security in the internet of things. IEEE Commun. Surv. Tutor. (2021). https://doi.org/10.1109/comst.2021.3067354

108. ur Rehman, M.H., Yaqoob, I., Salah, K., Imran, M., Jayaraman, P.P., Perera, C.: The role of big data analytics in industrial internet of things. Future Gener. Comput. Syst. **99**, 247–259 (2019). https://doi.org/10.1016/j.future.2019.04.020

109. Lei, S., Mithun, M., Michael, P., Noel, C., Son, N.H.: Challenges and research issues of data management in IoT for large-scale petrochemical plants. IEEE Syst. J. (2018). https://doi.org/10.1109/jsyst.2017.2700268

# Data Management, Privacy, and Trust
# in Distributed Systems

# Identity and Trust Architecture for Device Lifecycle Management

Agustín Marín Frutos[1]([✉])[ID], Jesús García Rodríguez[1][ID], Antonio Skarmeta[1][ID],
Konstantinos Loupos[2][ID], and Sokratis Vavilis[2][ID]

[1] Department of Information and Communication Engineering, University of Murcia,
30100 Murcia, Spain
{agustin.marinf,jesus.garcia15,skarmeta}@um.es
[2] INLECOM Innovation, Athens, Greece
{konstantinos.loupos,sokratis.vavilis}@inlecomsystems.com

**Abstract.** The Internet of Things (IoT) paradigm has become widespread, and only expected to increase in magnitude. The need for security and privacy measures adapted to this kind of environments has become apparent. This paper presents a comprehensive identity and trust architecture for managing the lifecycle of IoT devices, developed within the H2020 project ERATOSTHENES. The architecture focuses on secure device identity management, trust evaluations, and seamless interactions within multi-domain IoT environments. As part of the innovation, the solution includes the integration of self-sovereign identity principles and technologies such as Decentralized Identifiers, Verifiable Credentials and distributed ledger technologies. The proposed architecture leverages advanced cryptographic techniques and Physical Unclonable Functions to enhance security and privacy. The identity solution is complemented with a trust framework that continuously evaluates the domain's participants, establishing a zero-trust approach. This is enhanced with cyberthreat monitoring, sharing and mitigating tools to provide security across different planes. Thus, the framework addresses critical security and privacy requirements withing the lifecycle of devices, ensuring robust and scalable IoT device management from their initial bootstrapping to their decommissioning.

**Keywords:** IoT lifecycle · Identity management · Trust framework

## 1   Introduction

The Internet of Things (IoT) has revolutionised various sectors, from business and industry with Industry 4.0, to the daily lives of individuals through smart-home and medical technologies. This transformation has led to a surge in interconnected devices, each with distinct characteristics regarding computational power, memory, and energy consumption. The widespread presence of these devices, combined with their management of vast amounts of potentially sensitive data, underscores the growing importance of ensuring privacy and security within IoT ecosystems.

M. Presser et al. (Eds.): GIECS 2024, CCIS 2328, pp. 63–72, 2025.
https://doi.org/10.1007/978-3-031-78572-6_4

One crucial aspect to achieve this is the management of device identities. Each device needs a unique identity to ensure accurate tracking, communication and access control. Incorrect authentication of devices leads to security breaches, information leaks, and unauthorised access. Hence, establishing robust identity management protocols is fundamental to maintaining IoT environments.

The establishment of trust relationships is another of the main pillars at the core of IoT systems. This is necessary to ensure that devices can interact with each other and the rest of the system as well, while avoiding malicious behaviour. The trustworthiness of devices may depend on their behaviour, historical data, or compliance with security standards. By implementing trust management mechanisms, we can prevent malicious devices from compromising the network, detect their presence promptly, and inform relevant stakeholders to take appropriate action; ensuring the integrity of the data being exposed.

The security configuration of devices and their interaction with a domain is not a trivial process. Devices' lifecycle follows a complex process from deployment to decommissioning, going through various stages with different requirements. A secure deployment process, composed by the initialisation of the devices and bootstrapping and enrolment processes where trust and identity are established, is crucial to achieve security goals within a domain. Then continuous procedures of monitoring and updates are executed to keep the safety and trustworthiness of the environment and keep away malicious sources and vulnerabilities. Finally, when devices reach end of life, secure decommissioning ensures sensitive data is not left exposed.

The proliferation of Internet of Things (IoT) devices has necessitated robust frameworks for managing device identities and trust throughout their lifecycle. Solutions that address all these aspects will be useful for application to real-world scenarios. While some existing solutions tackle individual aspects (e.g., identity management within a security domain), they fail to comprehensively approach these tightly intertwined issues. This paper presents a detailed identity and trust architecture. The architecture aims to ensure secure and efficient management of device identities, trust evaluations, and secure interactions in multi-domain IoT environments, taking into account all the stages of devices' lifecycles.

The rest of the paper is organised as follows. Section 1.1 gives an overview of related work. Section 2 details the proposed architecture, detailing its components, functionalities and discussing some security, privacy and practical considerations. Lastly, Sect. 3 concludes the paper.

## 1.1    Related Work

The current landscape of IoT device management includes several approaches to identity and trust management. Existing solutions, such as GAIA-X [1] and FIDO's Device Onboard Specification (FDO) [6], provide foundational elements for secure identity management but often fall short in addressing the complete lifecycle and multi-domain interactions of IoT devices.

GAIA-X focuses on establishing a secure, federated data infrastructure for Europe, enabling secure and transparent data exchanges while ensuring compliance with European data protection laws [1]. However, GAIA-X primarily addresses data sovereignty and interoperability for cloud services rather than the specific needs of IoT devices and their lifecycle management. In contrast, the FIDO Device Onboard Specification aims to simplify and secure the initial onboarding of IoT devices but does not comprehensively cover ongoing lifecycle management or multi-domain trust interactions [6].

Moreover, the concept of self-sovereign identity (SSI) has gained traction as a means to provide users with control over their digital identities. This approach is supported by various standards and specifications, such as Decentralised Identifiers (DIDs) [7] and Verifiable Credentials [13]. These standards enable the creation of interoperable and privacy-preserving identity systems that can be integrated into IoT environments. Various studies propose the adoption of this approach in IoT environments [3,5,12], but fail to address specific IoT challenges like initial bootstrapping, privacy or highly varying trustworthiness conditions.

Various works have highlighted the importance of secure identity and trustworthiness management throughout devices' lifecycles. For instance, the work by Smith [18] provides a comprehensive guide to IoT security, emphasising the importance of robust identity management and trust frameworks. Additionally, Johnson and Doe [9] discuss device identity management in IoT networks, highlighting the challenges and potential solutions in this area.

Following the trend of decentralisation, various works study practical aspects of the application of decentralised technologies, particularly DLTs such as blockchain, to the IoT field as enablers for identity and trust management [4,11,15,17,19]. However, the scope of this works is limited to specific aspects, such as particular constraints of applying blockchain in IoT scenarios, or reduced frameworks for establishing trustworthiness. Our architecture leverages distributed ledger technologies (DLTs) and smart contracts to ensure the integrity and auditability of identity and trust data in a comprehensive way.

Recent advancements also include the development of risk-based automated assessment and testing frameworks for the cybersecurity certification and labelling of IoT devices [16]. These frameworks aim to provide a systematic approach to evaluating and improving the security posture of IoT devices and their incorporation to security domains. In this sense, the Manufacturer Usage Description (MUD) standard [10] is gaining traction. Multiple works propose advances over the standard, addressing issues such as its lack of expressiveness or extending them to threat modelling and mitigation [8,14]. Nonetheless, there remain challenges like establishing links between MUD files and device identities or domain's trust environments.

In summary, while existing solutions provide a solid foundation for addressing core aspects of secure IoT environments, there is a need for more comprehensive frameworks that address the entire device lifecycle and facilitate multi-domain interactions.

## 2   Proposed Architecture

The proposed architecture, developed in the context of the H2020 ERATOS-
THENES project,[1] is designed to facilitate a decentralised trust and identity
management framework for IoT environments. The architecture is structured
around multiple functional blocks, including device identity management, trust
evaluation mechanisms, and lifecycle management tools. The differentiation of
domains within the architecture allows for scalability and flexibility, enabling
tailored trust and identity management processes across diverse IoT scenarios.

The architecture is shown in Fig. 1 comprises several key components that
work in tandem to ensure secure and efficient management of IoT devices
throughout their lifecycle. These components include modules in each security domain's infrastructure, across the whole ecosystem, and within the devices
themselves.

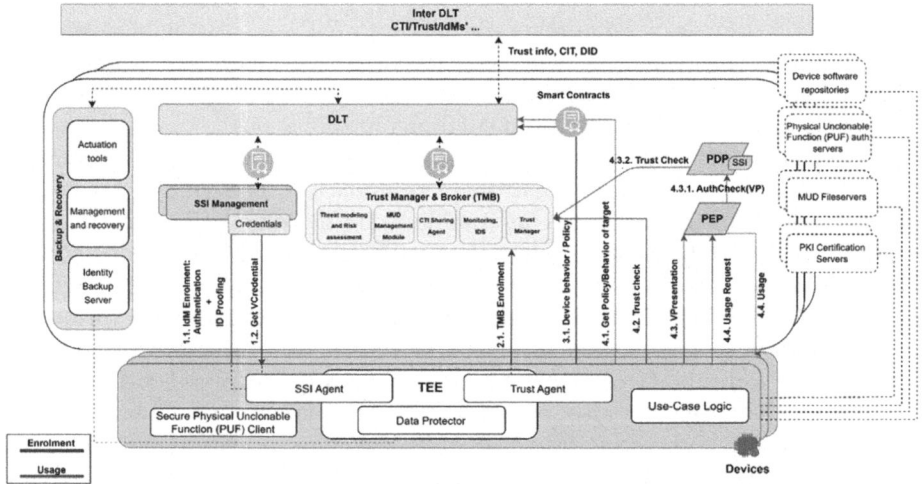

**Fig. 1.** Proposed architecture for identity, trust and lifecycle management

### 2.1   Components

The main components of a domain's infrastructure include:

– **SSI Management**: Manages device identities using self-sovereign principles,
  ensuring that devices can manage their identities during their operation in
  the domain without relying on a central authority. This component leverages
  decentralised identifiers (DIDs) and verifiable credentials (VCs) to provide
  secure and privacy-preserving identity management.

---

[1]  https://eratosthenes-project.eu/.

- **Trust Manager and Broker (TMB)**: Evaluates and manages trust scores for devices based on their behaviour and interactions within the domain. This component uses advanced risk assessment models and continuous monitoring to maintain an up-to-date trust score for each device. The TMB is further divided into several key sub-components, crucial in the security lifecycle of the devices. For instance, the CTI Sharing component is responsible of sharing of Cyber-Threat Intelligence (CTI) within and between domains, enhancing threat detection and response capabilities; while MUD Management component is in charge of disposing files of devices, describing security configurations and threat mitigation actions, to ensure secure and compliant device operations within the domain.
- **Backup and Recovery**: Handles device identity and data recovery processes, ensuring that devices can recover their identities and critical data in case of compromise or malfunction. This component works closely with the Data Protector module to securely store and retrieve sensitive data. Additionally, it supports stateful updates to ensure that devices can retain their trustworthiness and operational continuity through seamless transitions during software upgrades, e.g. as part of a response to a threat.
- **Authorisation Services**: Manages access control decisions and policy enforcement, ensuring that devices can only access services and resources for which they have been authorised. This component includes the Policy Decision Point (PDP) and Policy Enforcement Point (PEP) modules, establishing a zero-trust approach for continuous authorisation of devices.

To participate in the ecosystem and interact with the advanced functionalities provided, devices need specific components:

- **PUF Authentication client:** This module will be available on PUF-enabled devices, and will be in charge of the PUF-based cryptographic operations, acting as a root of trust for the device's identity.
- **TEE:** The Trusted Execution Environment will act a as supporting element for confidential computations, enabling secure management of sensitive material and execution of trustworthy code.
- **SSI Agent:** The agent will manage the interactions of the device with the SSI framework. It deals with keys and attribute-based credentials to enable authentication and authorisation processes in the domain.
- **Trust Agent:** It handles multiple activities related to device trust management, such as monitoring and the interactions with the Trust Manager and Broker.
- **Data Protector:** It handles data that has to be stored securely and provides secure backup and recovery mechanisms.

Lastly, some components will be transversal to the whole ecosystem, as manufacturer services enabling advanced capabilities:

- **Device Software Repositories**: Provide access to software and firmware, enabling deployment and updates for devices in the ecosystem.

- **PUF Authentication Servers**: Provides a secure method for authenticating devices using Physical Unclonable Functions (PUFs). These servers generate unique cryptographic keys based on the inherent physical characteristics of each device, making the keys highly resistant to cloning and tampering. This authentication method significantly enhances the security of device identity verification processes.
- **MUD Fileservers**: Manages and distributes Manufacturer Usage Description (MUD) files to ensure devices operate within their expected parameters. These servers store and provide access to MUD files that define behaviour and security policies for devices, for instance limiting device communications to only those specified by the manufacturer. MUD Fileservers handle both standard MUD files and threat MUD files, which include additional security measures to respond to identified threats. The distribution process uses secure protocols to maintain the integrity and authenticity of the MUD files during retrieval and distribution.

## 2.2   Functionalities and Processes

The Internet of Things (IoT) ecosystem presents unique challenges and opportunities, necessitating comprehensive solutions that address the entire lifecycle of IoT devices. The architecture proposed in the paper, developed within the ERATOSTHENES project, ensures secure and efficient management of IoT devices, encompassing their initialisation, authentication, service usage, threat monitoring, assessment, sharing, as well as update and recovery processes. This is articulated through a series of processes, each aimed at reinforcing the security and functionality of IoT environments. Among them, the most representative are:

- **Bootstrapping and Enrolment**: Initial device configuration, identity registration, and trust evaluation in the target domain. This process involves the configuration of the device's PUF, retrieval of domain-specific metadata, and enrolment of the device in the domain's identity and trust management systems. The authentication steps involving PUF and identity proofing are closely linked with the enrolment process to establish the device's final identity in the domain. This linkage ensures that the device's identity is verified and trusted before it is enrolled in the domain. Additionally, MUD (Manufacturer Usage Description) files are utilised to configure the security policies for the device within the domain, ensuring compliance with the domain's security requirements. At the end of this process, the device will hold verifiable credentials that represent its identity in the domain, will be registered in the trust framework with an initial trust score, and be able to operate in the domain.
- **Service Usage**: Authorisation and access to domain-specific services. Devices generate verifiable presentations from the credentials obtained during enrolment to prove their identity and trustworthiness, which are then validated by the domain's authorisation infrastructure. Privacy is maintained

through the use of Attribute-Based Credentials (ABCs), allowing for fine-grained access control policies. The domain employs a zero-trust approach, utilising PDP (Policy Decision Point) and PEP (Policy Enforcement Point) to ensure continuous validation of device credentials and trust level. Trustworthiness measures are backed by Distributed Ledger Technology (DLT), providing a tamper-proof and transparent record of trust evaluations and authorisation decisions.

– **Threat Assessment**: Continuous monitoring and evaluation of device behaviour to manage trust scores, according to models of the device and its role in the domain's context. The Trust Manager and Broker (TMB) module collects and analyses data from various sources to update the risk and trust scores of devices.
– **Cyber-Threat Intelligence Sharing**: Secure sharing of threat information across domains. The CTI Sharing Agent facilitates the exchange of Cyber-Threat Intelligence (CTI) reports between domains using privacy-preserving techniques. In line with NIS2 directive [2], this project enables inter-DLT sharing of CTI and threat data across domains and with the whole ecosystem, including CSIRTs. The use of immutable DLT technologies ensures auditability, while privacy means like pseudonymisation techniques are applied.
– **Secure Software Deployment**: Ensures integrity and authenticity of software updates. The Management and Recovery component coordinates the deployment of software updates, verifying their integrity and authenticity before installation. This process can be triggered by mitigation actions mandated by the MUD Management or CTI sharing components in the TMB. MUD files provide recommendations and constraints on device behaviour, which are used to generate access control lists and enforce security policies. The CTI sharing agent collaborates with TMB to share threat intelligence across domains, facilitating the detection and mitigation of security threats. The TMB continuously monitors device behaviour and trust levels, using this data to assess risks and trigger necessary software updates to mitigate identified vulnerabilities. This integrated approach ensures that the software deployment process is both secure and responsive to emerging threats.
– **Identity Recovery**: Processes for recovering device identities in case of compromise or malfunction. The Identity Backup Server and Data Protector modules work together to securely restore device identities and critical data.

### 2.3 Discussion

The architecture addresses several critical security and privacy requirements to ensure the robustness and reliability of the identity and trust framework. Key considerations include:

– **Protection against Trust Management Attacks**: The use of decentralised ledger technologies (DLTs) and smart contracts ensures that trust evaluations and modifications are secure, auditable, and tamper-proof. By storing trust scores and the rationale for trust modifications on a DLT, the architecture prevents unauthorised alterations and enhances transparency.

- **Data Minimisation and Privacy Compliance**: The architecture adheres to GDPR requirements by implementing data minimisation techniques and providing mechanisms for data management. Components that manage identity information ensure compliance with GDPR and other relevant privacy regulations.
- **Root of Trust for Identification**: The use of Physical Unclonable Functions (PUFs) provides a robust method for authenticating devices, ensuring that each device has a unique, tamper-resistant identity. PUFs generate cryptographic keys based on the physical properties of the device, making them highly resistant to cloning and forgery.
- **Intrusion Detection, Monitoring and Threat Reporting**: The architecture includes intrusion detection systems (IDS) and continuous monitoring tools to detect and respond to security threats in real-time. These components work in tandem with the CTI Sharing Agent to share threat intelligence and enhance the overall security posture of the ecosystem.
- **Secure Software and Firmware Updates**: The architecture ensures the integrity and authenticity of software and firmware updates through the use of cryptographic techniques. The Device Software Repositories component verifies the authenticity of updates before deployment, and devices only accept signed and validated software packages.
- **Secure Adaptability**: While the full functionality is described above, the framework is designed with flexibility in mind to cope with the heterogeneity of devices. For instance, devices may not be PUF-enabled, or provide different levels of hardness for the Trusted Execution Environment such as hardware versus software implementations. This will be possible and detected during the enrolment phase. Thus, the trust framework will be able to take the capabilities into account for its evaluation, enabling the heterogeneous environment while ensuring security.
- **Interoperability**: The architecture is aligned with key frameworks to enhance interoperability of the developed solutions. For instance, the identity solution relies on **Decentralised Identifiers (DIDs)**, that facilitate secure and verifiable device and infrastructure identities. DIDs provide a standardised method for devices to interact securely and privately within the ecosystem. For fine-grained authorisation, the architecture uses **Verifiable Credentials (VCs)**, ensuring secure and privacy-preserving attribute-based authentication. The use of VCs along with ABCs enables devices to prove specific attributes or qualifications without revealing unnecessary information, adhering to the principles of minimal disclosure while adhering to current standards in self-sovereign solutions.

## 3   Conclusion

The rise of the IoT paradigm requires solutions that comprehensively address the various specific challenges of such systems. This paper has presented a detailed identity and trust architecture for managing the lifecycle of IoT devices. The intertwined integration of advanced identity management techniques, trust evaluation mechanisms, and lifecycle management tools provides a comprehensive solution for secure IoT device management in multi-domain environments. By leveraging state-of-the-art technologies such as DIDs, VCs, PUFs, and DLTs, the proposed framework ensures robust security, privacy, and scalability for IoT ecosystems.

Future work will focus on the practical implementation and validation of the framework in real-world scenarios, exploring its applicability across various IoT domains and use cases. Additionally, ongoing research will aim to refine the components and processes within the framework to address emerging security and privacy challenges in the ever-evolving IoT landscape, and linked to emerging initiatives like data spaces.

**Acknowledgments.** The research leading to this work has been partially funded by the European Union's Horizon 2020 research and innovation program under Grant Agreement No 883335 (ERATOSTHENES).

**Disclosure of Interests.** The authors have no competing interests to declare that are relevant to the content of this article.

# References

1. Gaia-x technical architecture (2020). https://www.data-infrastructure.eu/GAIAX/Redaktion/EN/Publications/gaia-x-technical-architecture.pdf
2. Directive (eu) 2022/2555 of the european parliament and the council on measures for a high common level of cybersecurity across the union (2022). https://eur-lex.europa.eu/eli/dir/2022/2555/oj
3. Bouras, M.A., Lu, Q., Dhelim, S., Ning, H.: A lightweight blockchain-based iot identity management approach. Future Internet **13**(2), 24 (2021). https://doi.org/10.3390/fi13020024
4. Chen, X., Nakada, R., Nguyen, K., Sekiya, H.: A comparison of distributed ledger technologies in IoT: IOTA versus Ethereum. In: 2021 20th International Symposium on Communications and Information Technologies (ISCIT), pp. 182–187. IEEE, Tottori, Japan (October 2021). https://doi.org/10.1109/ISCIT52804.2021.9590601, https://ieeexplore.ieee.org/document/9590601/
5. Cocco, L., Tonelli, R., Marchesi, M.: A system proposal for information management in building sector based on BIM, SSI, IoT and blockchain. Future Internet **14**(5), 140 (2022).https://doi.org/10.3390/fi14050140
6. Cooper, G., et al.: Fido device onboard specification 1.1 (2021)
7. Drummond Reed, Manu Sporny, Dave Longley, and Jonathan Holt: Decentralized identifiers (dids) v1.0 (2020). https://www.w3.org/TR/did-core/
8. Heeb, Z., Kalinagac, O., Soussi, W., Gür, G.: The impact of manufacturer usage description (mud) on iot security. In: 2022 1st International Conference on 6G Networking (6GNet), pp. 1–4. IEEE (2022)
9. Johnson, M., Doe, J.: Device identity management in iot networks. In: Proceedings of the 2019 International Conference on Internet of Things (ICIoT), pp. 75–80. IEEE (2019)

10. Lear, E., Droms, R., Romascanu, D.: Manufacturer usage description specification. Technical report, RFC Editor (2019)
11. Liu, Y., Zhang, C., Yan, Y., Zhou, X., Tian, Z., Zhang, J.: A semi-centralized trust management model based on blockchain for data exchange in iot system. IEEE Trans. Serv. Comput. **16**(2), 858–871 (2022). https://doi.org/10.1109/TSC.2022.3181668
12. Lücking, M., Fries, C., Lamberti, R., Stork, W.: Decentralized identity and trust management framework for internet of things. In: IEEE International Conference on Blockchain and Cryptocurrency, ICBC 2020, Toronto, ON, Canada, 2–6 May 2020, pp. 1–9. IEEE (2020).https://doi.org/10.1109/ICBC48266.2020.9169411
13. Sporny, M ., Longley, D., Chadwick, D.: Verifiable credentials data model v1.0 (2019). https://www.w3.org/TR/vc-data-model
14. Matheu García, S.N., Sánchez-Cabrera, A., Schiavone, E., Skarmeta, A.: Integrating the manufacturer usage description standard in the modelling of cyber-physical systems. Comput. Stand. Interfaces **87**, 103777 (2024)
15. Popa, M., Stoklossa, S.M., Mazumdar, S.: ChainDiscipline - Towards a Blockchain-IoT-Based Self-Sovereign Identity Management Framework. IEEE Trans. Serv. Comput. **16**(5), 3238–3251 (2023). https://doi.org/10.1109/TSC.2023.3279871, https://ieeexplore.ieee.org/document/10135155/
16. Matheu-Garcia, S.N., Hernandez-Ramos, J.L., Skarmeta, A.F., Baldini, G.: Risk-based automated assessment and testing for the cybersecurity certification and labelling of iot devices. Comput. Stand. Interfaces **62**, 64–83 (2019)
17. Saeed, M., Amin, R., Aftab, M., Ahmed, N.: Trust management technique using blockchain in smart building. Eng. Proc. **20**(1), 24 (2022). https://doi.org/10.3390/ecsa-9-12810, https://www.mdpi.com/2673-4591/20/1/24
18. Smith, J.: IoT Security: A Comprehensive Guide. Tech Publishers, Bangalore (2020)
19. Zhang, K., Lee, C.K.M., Tsang, Y.P.: Stateless blockchain-based lightweight identity management architecture for industrial IoT applications. IEEE Trans. Ind. Inform. 1–12 (2024). https://doi.org/10.1109/TII.2024.3367364, https://ieeexplore.ieee.org/document/10468559/

# Privacy Preserving Enablers for Data Space Ecosystems

Natalia Borgoñós García[✉], María Hernández Padilla,
and Antonio Fernando Skarmeta Gómez

University of Murcia, Murcia, Spain
natalia.borgonosg@um.es

**Abstract.** Data Spaces are ecosystems designed to allow multiple organizations or companies to share data in a secure manner. Despite the potential of these technologies, they encounter a number of challenges and privacy issues that limit their use. Privacy Preserving Enablers are mechanisms developed to tackle these difficulties, ensuring data integrity and access control. This paper aims to analyze the role of some Privacy Preserving Enablers and its integration with Connectors in the context of Data Spaces. The research will focus on key enablers, including a Self-Sovereign Identity with Zero-Knowledge Proof, which is a privacy preserving approach that allows users to verify their identity and attributes without the need to disclose underlying data, ensuring their privacy. Additionally, the usage of Sticky Policies instantiated through Attribute-Based Encryption attaches control policies into the encrypted data in order to have an attribute-based access control, enhancing its security. The application of Policies Enforcement assure the consistent application of policies and the maintenance of the security within the Data Space.

**Keywords:** Data Spaces · Privacy Preserving Enablers ·
Self-Sovereign Identity · Sticky Policies · Policy Enforcement

## 1 Introduction

Nowadays, data has become an essential asset for every organization. The increasing importance of data has led to the creation and development of Data Spaces [1], that are ecosystems where organizations are able to share data coming from different sources and collaborate together to achieve a goal. The emergence of Data Spaces represents a shift to the way that companies and organizations share and manage data. It presents a collaborative approach, where a secured and structured environment for data usage. The use of Data Spaces encourages the cooperation between the parts involved, recognizing the need for effective communication that this process requires.

Nonetheless, the use of Data Spaces also comes with some challenges, particularly considering the need to preserve data protection and privacy within these ecosystems. Since data is being shared among multiple parts, a single error can

© The Author(s) 2025
M. Presser et al. (Eds.): GIECS 2024, CCIS 2328, pp. 73–88, 2025.
https://doi.org/10.1007/978-3-031-78572-6_5

compromise the entire system's functionality. Common risks within the scope of Data Spaces include data breaches, privilege abuse, data theft or data tampering. These problems may arise from unauthorized access or attacks that affect the integrity and authenticity of the data. Such breaches could lead to misuse or unauthorized viewing of sensitive information by individuals outside of the organization or those not allowed to access the data.

As Data Spaces continue to evolve, there is an absence of proper tools, mechanisms and technologies to ensure data privacy and integrity. Despite the potential of Data Spaces, these issues limit their adoption, given that organizations seek an environment where their data can be securely maintained. This motivates the development of Privacy-Preserving Enablers, that claim to mitigate the risks that come with the usage of Data Spaces. Privacy-Preserving Enablers are methodologies designed to protect the data confidentiality during its use. This article aims to present some Privacy-Preserving Enablers to propose a solution for the problems that Data Spaces present, thereby enhancing trustworthiness and security.

Bringing about this study about Privacy-Preserving Enablers contributes to advancing security in Data Spaces by fostering secure data-sharing ecosystems. Not only do these technologies ensure data full protection within the Data Space, but they also guarantee private and protected collaboration among the involved participants. The development of privacy-preserving measures in the field of Data Spaces give rise to the growth and usage of these ecosystems. This article seeks to promote and supply insights for effective Data Spaces and awareness of having a suitable privacy approach to data itself. This underscores the essential role of the Privacy-Preserving Enablers in advancing Data Space ecosystems. To achieve this objective, the article will draw on previous work to demonstrate how Privacy-Preserving address the challenges that come with Data Spaces.

The document is organized as follows:

**Section 2: Related Work.** This section provides an overview of privacy-preserving tools and techniques relevant to Data Spaces and discusses the application in other contexts.

**Section 3: Data Space Ecosystems.** Introduction to Data Space concept, its architecture and gap analysis.

**Section 4: Privacy-Preserving Enablers Definition.** Description of the Privacy-Preserving Enablers introduced and their role in addressing the challenges as well as an evaluation on how they integrate within Data Space Connectors.

**Section 5: Conclusions and Future Work.** The document concludes with a summary of the points discussed, the importance of Privacy-Preserving Enablers and its contributions in the evolution of Data Spaces. It also discusses potential future research directions.

# 2 Related Work

This section will provide an overview of the state-of-the-art techniques for preserving privacy and security within Data Spaces. The main Privacy-Preserving Enablers provided by will be introduced here, highlighting their background and comparing them to other tools used for similar purposes.

## 2.1 Identity Management

Identity Management is a crucial element that verifies the identity of the entities that take part in a network or an environment. Recently, there has been an increase in the number of participants within ecosystems, so robust identity management solutions are essential for preserving privacy and trustworthiness within Data Spaces.

Centralized Identity Management represents the traditional approach in which a single provider manages all user identities, such as LDAP-based systems [2]. The main drawback of this traditional model is its centralized nature, relying on one third party to authenticate user identities for data access. This creates a single point of failure introducing privacy concerns. Attempts to address the limitations of centralized identity management include Federated Identity Management, which enables users to authenticate across several systems using a single set of credentials. While OpenID Connect (OIDC) [3] is a standard that supports this approach, it continues to face challenges [4] related to privacy and trustworthiness dependence on other entities.

Finally, the emergence of Self-Sovereign Identity (SSI) has involved a major change in Identity Management systems, although the adoption of this approach is still in its early stages, it is able to be integrated with traditional systems that have been previously mentioned through specific integration patterns that provide a promising solution to overcome the limitations of centralized identity approaches [5]. Users have full control over their identity without relying on centralized authorities, managing their identity by presenting proofs without revealing non-essential information. This approach enhances privacy and security in the context of Data Spaces. The proofs used by users to authenticate themselves can be based on a cryptographic technique called Zero-Knowledge Proof (ZKP) that have been developed and improved in recent years [6]. ZKP enables the verification of user attributes without disclosing unnecessary information. Using this mechanisms the security can be enhanced because the main goal could be achieved maintaining privacy.

## 2.2 Confidentiality and Privacy by Design

The real value of data stem from its usage and processing, which underscores the need to adopt a data-centric approach that comprehensively addresses privacy and security issues. As mentioned above, users should have sovereignty over their data and the ability to choose the disclosed information. In this point, it is crucial to provide users with tools to control their data and the way they are shared. The principles of Privacy by Design [7] are a fundamental notion, where proactive protection, privacy embedded into design and lifecycle protection of data play a basic role.

There are some challenges that need to be tackled, like the lack of balance in data privacy and utility or user control over data access. The usage of anonymization techniques [8], encryption and cryptographic schemes and Attribute-Based Encryption (ABE) can be used to deal with these issues. ABE is a mechanism for encrypting data where access control policies are defined based on user attributes, thereby enhancing fine-grained access control and ensuring that only authorized consumers can access the data.

Looking into these approaches to solve the privacy problems, Functional Encryption and Sticky Policies are interesting mechanisms. Functional Encryption enables decryption modules to learn a function of the plaintext, whereas Sticky Policies are policies instantiated through ABE and attached to the encrypted data that let users to maintain control over their data. These policies can be easily integrated with ABE and enforced, encrypted and set depending on the context [9].

## 3   Data Spaces Ecosystems

The concept of Data Spaces, along with its motivation and challenges have been previously introduced [10]. However, its architecture and relevant concepts require further explanation. The design of the architecture of Data Spaces is focused on interoperability and security, counting on several basic roles to ensure these qualities, represented in Fig. 1 [11].

**Data Providers** are entities that offer data through the Data Space catalogue. The rights to access and use this data are regulated by **Data Owners**.

**Data Consumers** access the data provided, while **Data Intermediaries** are services that are offered to facilitate and offer the access of data. They preserve the catalogue of organised data with information, use application tools to control quality or operational aspects.

**Technology providers** are entities that work enabling the utility of the Data Space, while **Operators** will focus on the proper definition, management and maintenance of the Data Space.

A Data Space could count on different components depending on the approach adopted [12]. Despite of this, there are some of them that are strictly necessary for a Data Space to perform its basic activity as required.

One of these components and one of the main elements in the context of data spaces is a **Connector**. This component acts as a mediator between Data Providers and Consumers in the Data Space [13], facilitating secure data exchange between them by applying the usage of mechanisms that enable the protection and privacy to keep data safe. Some of these mechanisms are known as the Privacy Preserving Enablers that are being presented in this document.

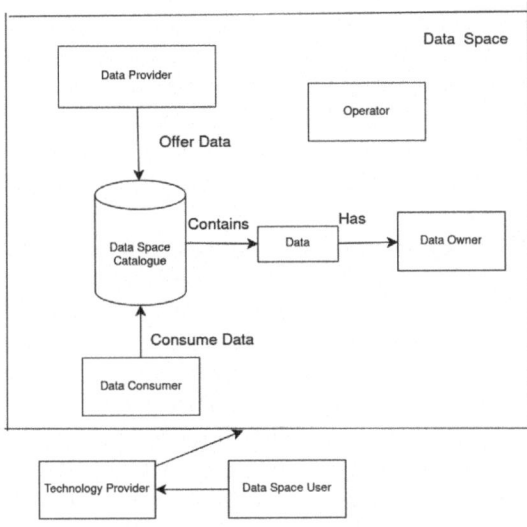

**Fig. 1.** Data Space basic roles.

In the context of operating within a data space environment, there are several key initiatives and frameworks which share a common goal of fostering innovation, compliance with regulations like GDPR [14], collaboration and digital sovereignty within European's digital ecosystem- some of them will be reference throughout the paper as every development aims to align with them. These initiatives include, among others, IDSA [15], Gaia-X [16], FIWARE [17], EOSC [18], DSSC [19] and DSBA [20]. These frameworks provide essential guidelines and standards for implementing, secure, interoperable data practices across organizations.

After explaining the keys to undertake a safe Data Space, the Privacy Preserving Enablers introduced before are going to be deeply treated and associated with Data Spaces.

## 4   Privacy Preserving Enablers Definition

In our highly connected digital world, where data is increasingly exchanged across multiple platforms and services, ensuring privacy and security of sensitive information has gained great importance. Privacy-preserving technologies play a crucial role in protecting personal and confidential data from unauthorized access, misuse, and breaches. This is particularly critical in data space connectors, which facilitate the secure and efficient exchange and management of data across different organizations and systems. There are several reasons why these technologies are important for data space connectors:

- Compliance and Regulation: privacy-preserving technical enablers are crucial for organizations to comply with data protection regulations such as GDPR and the proposed Data Act and Data Governance Act [21]. In the context of data space connectors, DSSC describes a Blueprint [22] guideline where a Governance and Trust framework is defined for ensuring its secure, compliant and efficient use. Mechanisms like encryption or strict access controls are required by such regulations.
- Trust Management: organizations demonstrate their commitment to protecting user privacy by implementing privacy-preserving tools. The Trust Framework focuses on building and maintaining trust among stakeholders in a data ecosystem, protecting data from unauthorized access, breaches or tampering. Key mechanisms for reaching such a purpose are authentication and authorization, encryption and Self-Sovereign Identity (SSI).
- Data Protection: the risk of sensitive information exposure through data exchange and storage is mitigated using such technologies- only authorized entities have access to specific data based on predefined policies and user permissions. Apart from the defined data space Trust Framework, the Governance one is involved here- policies and procedures are established for managing data throughout its life-cycle. Key mechanisms are policies definition and policy enforcement.

Several methods to achieve the above aims and requirements have been explored and are put forward in this paper.

### 4.1   SSI Solution with Zero-Knowledge Proof

One of the areas within data connectors that requires heightened privacy and confidentiality is identity (trust) management. Identity information involves sensitive data which, if compromised, can lead to significant data breaches and privacy violations. The use of a privacy-preserving and self-sovereign identity (SSI) approach empowers end-users with more control over their own personal data without the need to rely on a central authority. There are three main participants in an SSI solution: holder of Verifiable Credentials (VCs) with a digital wallet to store and present VCs; issuer of VCs, and verifier who checks VCs. Additionally, a Data Registry (e.g., Blockchain), which stores the information in a decentralized way, is needed.

To align with the concept of data spaces and the DSBA Technical Convergence [23], where one of the primary goals is to work on common models to create these spaces, the SSI proposal will adhere to various common international and European standards for identity management:

- Strong alignment with international standards and EU initiatives (W3C [24], OpenID [25], EBSI-ESSIF [26])
- W3C Verifiable Credentials data Model v2.0 [27]
- W3C Decentralized Identifiers v1.0 [28]
- Self-Issued OpenID Provider (SIOP) v2 [29]

- OpenID for Verifiable Presentations (OID4VP) [30]
- Verifiable Credential Issuance (OID4VCI) [31]
- Enhanced GDPR compliance (e.g. ZKP)

In addition to the basic SSI model stated above, a cryptographic module [32] including distributed attribute-based credentials leveraging Pointcheval-Sanders multisignatures [33] (dp-ABC) and zero-knowledge proofs(ZKP) is incorporated in the SSI components to offer a robust and comprehensive approach to privacy-preserving identity management. Dp-ABC credentials allow users to disclose only the necessary attributes required for specific transactions (Selective Disclosure), ensuring that sensitive information remains protected and that users have greater control over their personal data. ZKPS are generated for the disclosed attributes, enabling the prover to demonstrate to the verifier that a statement is true (they fulfill the specific attributes) without revealing the underlying data. Thanks to this cryptographic module, the security and privacy of the Self-Sovereign Identity (SSI) mechanisms are strengthened, giving users greater control over their personal data.

The cryptographic primitive of this module is integrated into W3C VCs. Verifiable Credentials will be generated and signed by an issuer, and the holder will have complete control over them from that point. Holders will carry out authentication processes directly against verifiers. For that, they can use their VCs to derive Verifiable Presentations, which are a tamper-evident way to gather and share identity information from the credentials for a presentation process.

In the provided solution, ZKP will be used to modify the signed VC so that only part of the information is revealed, while keeping the formal authenticity guarantees. Thus, the Verifiable Presentation will contain the derived credential, increasing the privacy of the holder.

Additionally, the solution presented in this paper uses as reference the open-source wallet of walt.id [34], which aims to create a shared, reusable, interoperable tool kit designed for initiatives and solutions focused on creating, transmitting and storing verifiable digital credential, following SIOP and OIDC4VP specifications. The functionalMARIA HERNANDEZ PADILLAities provided by walt.id solution is enriched with dp-ABCs and ZKP mechanisms along with additional modifications to meet all specified standards.

Two different flows (Fig. 2, Fig. 3) are defined for issuing and verifying a credential using ZKPs:

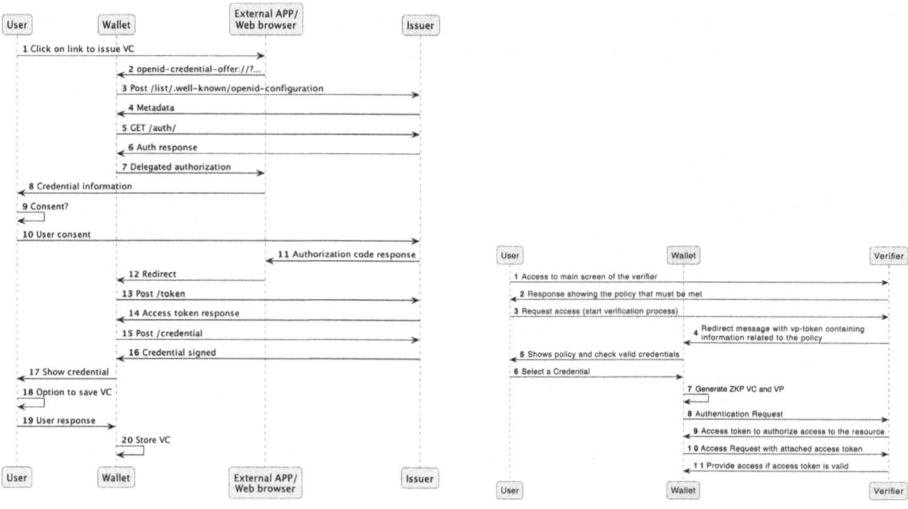

**Fig. 2.** Credential Issuance flow.          **Fig. 3.** Credential Verification flow.

## 4.2 Sticky Policies Instantiated Through Attribute-Based Encryption (ABE)

In the context of Data Spaces, sticky policies policies instantiated through Attribute-Based-Encryption enable a fine-grained access control based on attributes. This is a robust mechanism to ensure data privacy, that data is only accessible to authorized members within the Data Space and its compliance with security conditions.

To design these Sticky Policies, the main steps that might be followed are:

1. **Definition.** This step focuses on user or device attributes that may be significant to the Data Space. It could be a role, a department or a specific permission. The policy could is defined in order to dictate the conditions that the data consumer must satisfy to access them. So, the consumer's attributes must fulfill the attributes that the policy state.
2. **Embedding.** The policy that has been defined in the previous step has to be embedded into the data that it protects, this will be achieved using ABE. Using the attributes that have been used will be employed to derive the encryption keys. Only the ones with matching attributes would be able to decrypt the data.
3. **Enforcement.** The attributes of an user of the Data Space that attempts to access the data will be evaluated and checked if they comply the embedded policies. Access will only be granted if his attributes match with the policy ones.

To deploy this functionality, data owners and administrators require a module for creating, managing sticky policies and integrating them into the data space. This module also is responsible of ABE-based encryption and decryption that checks user attributes against policies, with the power of ensuring policy fulfillment and granting access to the data.

The ABE Module counts on a Server side and a Client side. The Server side is where the tasks like policy management, encryption, decryption and data protection is carried out. The Client side is where the consumer starts the process of publishing or retrieving data, supplying the necessary information to finish the task properly.

The ABE Module Initialization, defined in Fig. 4, consists in:

1. The Server gets started and creates a keypair for a consumer. The public key goes to the General Infrastructure, associated with an *uuid*. (Steps 1–4)
2. The consumer uses the Client Side to present itself to the SSI Agent with its *uuid* and its attribute list. This originates a Verifiable Presentation (VP) that attaches to the previous information to request the encryption key to the Server to access the data. (Steps 5–10)
3. After the verification of the information received, and if everything seems to be correct, the server sends the key to the consumer and he stores it safely. (Steps 11–15)

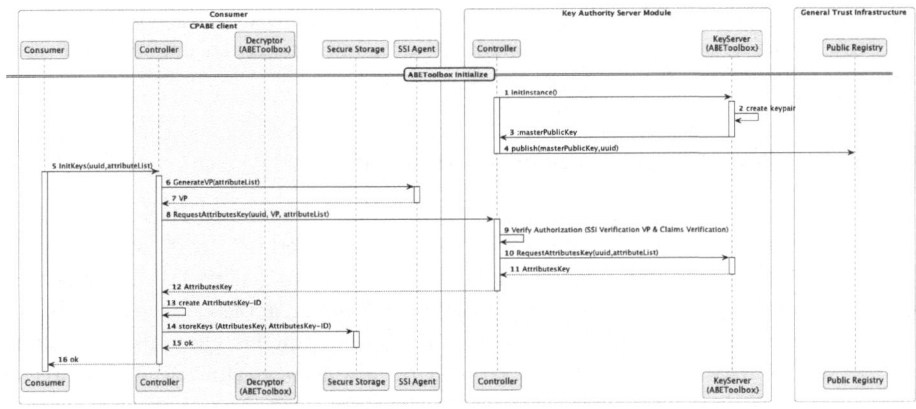

**Fig. 4.** ABE Module Initialization and Attributes key Generation.

For publishing data encrypted based on ABE, Fig. 5, the data producer requests the Client Side to encrypt data applying a certain policy, associated with its *uuid*. An Encryptor instance of the module will encrypt the data properly and return to the producer the encrypted data, that will be published in the cloud storage. (Steps 1–6)

For reaching decrypted data, the consumer requests the data to the storage. Then, the client side of the consumer receives a request to decrypt the attached encrypted data and an attribute list. The client retrieves the attribute keys that have been saved in the initialization and sends a request to the decryptor, that will check if the consumer's keys allows him to get the decrypted data. (Steps 7–13)

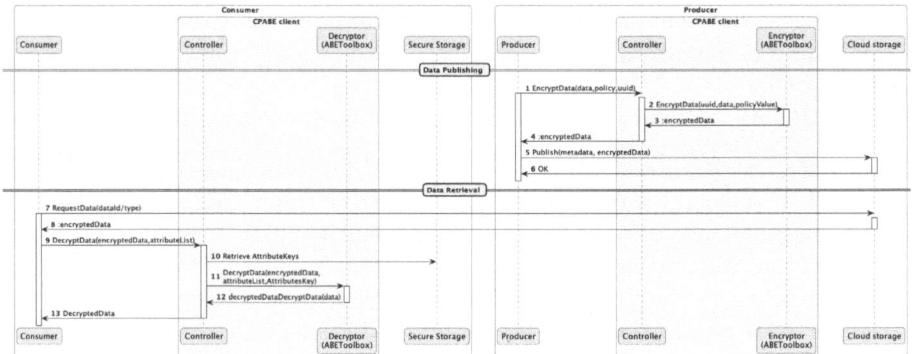

**Fig. 5.** ABE Module Data Publishing and Retrieval.

The intervention of this module is highly beneficial within the Data Spaces technologies as it gives a solution to the challenges that these ecosystems present in terms of preserving privacy. It provides efficient means to ensure that only authorized users are able to interact with data.

### 4.3   Access and Usage Policies Enforcement

Following the definitions in DSBA Technical Convergence and IDS-RAM4 [35], a policy enforcement model has been implemented based on the XACML framework [36] which, combined with sticky policies, create a secure layer of resource access authorization. User consent is addressed using access control based on attribute-based policies. XACML's attribute-based policies can integrate identity attributes which are associated to user preferences, permissions, or consent choices.

Once the authentication, as described in Sect. 2.1, has been performed, the verifier party issues an access token which contains the VCs needed to access protected resources.

The solution implements the following components:

- **Policy Enforcement Point (PEP):** It serves as the entry point of the enforcement process, which receives the access token obtaining the access request and forwards it to the PDP. Once the PDP has made a decision on whether to grant or deny access to the resource, PEP receives it back and either allows or locks the data according to decision.
- **Policy Decision Point (PDP):** Evaluates and makes a decision based on the access control policies and the access token provided by the PEP. The

result is forwarded to the PEP. Furthermore, PDP considers additional information about the context, not present in the request itself.

- **Policy Administration Point (PAP):** This is the instance that will manage and storage the whole information of the policies. The PDP requests this component the proper policy or policies that it needs in order to make a decision about granting access.
- **Policy Information Point (PIP):** Apart from the policy, the PDP could require further information to make the decision. This PIP storages some other information like trust scores that will help the PDP to know if the requester is suitable to receive the access permission.

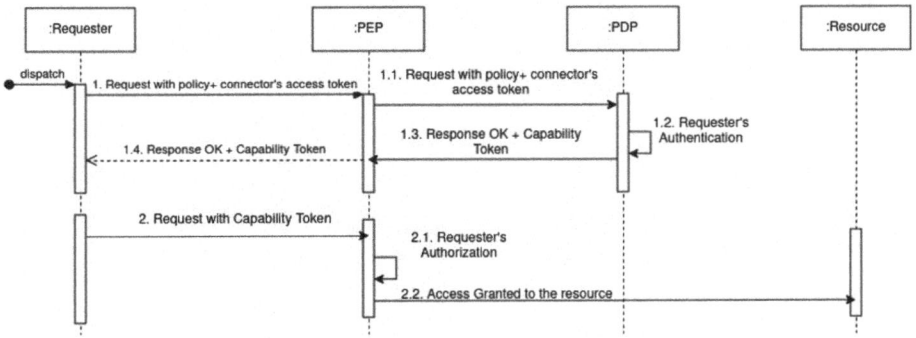

**Fig. 6.** Policy Enforcement workflow.

Figure 6 shows the flow and how the components interact with each other. The first part of the flow demonstrate the process of getting the access to the resource. When the requester achieves the access to the resource because everything is correct, he receives a Capability Token. This Capability Token will be presented to the PEP, that will verify that is correctly formed and signed, to show that the requester is allowed to access the resource. This Capability Token can only be used during a period of time, also checked by the PEP when the requester uses it.

### 4.4 Integration in Connectors

Considering the context and architecture of Data Space Connectors defined in Sect. 3, there are two main integration points in relation to the three different assets described in the preceding Sects. 4.1, 4.2 and 4.3.

SSI model integrates within the Identity Management of the Data Space: each participant, associated to one connector, must have a SSI wallet as the fundamental element of decentralized identity for DIDs, VCs and ZKPs management. Additionally, the data space infrastructure necessitates at least one VCs issuer and verifier.

Both sticky policies and policy enforcement mechanisms are addressed to access control, complementary to each other, which significantly enhance data security. Policy enforcement mechanisms enforce the defined policies across the distributed environment, safeguarding data integrity and confidentiality. Sticky policies ensure that access permissions remain attached to the data itself thanks to CP-ABE, providing a double layer of authorization and enabling granular control over who can access specific resources and under what conditions.

The following diagram (Fig. 7) describes the process of gaining access to a specific resource from the consumer side:

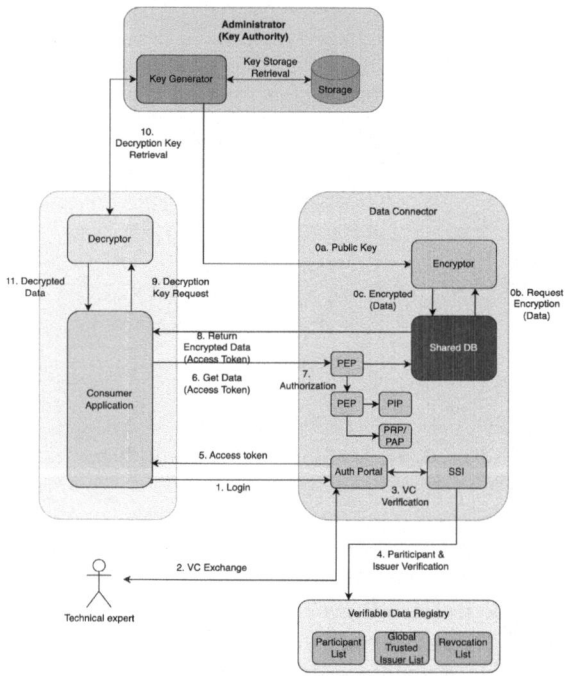

**Fig. 7.** Access Control workflow including sticky policies and policy enforcement.

This process contemplates the privacy-preserving enablers that have been mentioned along the document working together to achieve a consistent access control system. Its implementation have been tested and adapted in several occasions due to the usage of this component in two European projects, TANGO [38] and TITAN [39].

TANGO Project provides a secure, trustworthy and environmentally sustainable data management ecosystem to supply a high impact solution within several domains like transport, e-commerce, finance, public administration or tourism supporting numerous beneficiaries across Europe.

TITAN Project platform solution for confidential, secure and privacy-preserving data processing and collaboration.

The connector and privacy enablers that use these projects counts on several testing modules and scripts that prove its utility and functionality, integrated with the main tasks and activities of the system.

## 5 Conclusions and Future Work

Leveraging CP-ABE as sticky policies in conjunction with PEP/PDP architecture provides a robust framework for achieving secuocasiones inglesre, compliant, and flexible data access control in data space environments. Additionally, incorporating Self-Sovereign Identity enhanced with zero-knowledge proof increases privacy by enabling identity verification without disclosing sensitive information. This is particularly crucial within the European digital ecosystem, where initiatives such as IDSA, Gaia-X and EOSC aim to create references and best practices for data sharing environments across Europe.

Access control approached by sticky policies and PEP/PDP policy enforcement provide a fine-grained access control allowing data owners to define by means of attribute-based policies who can access their data and on what conditions. In addition, data sovereignty is enhanced ensuring that data access is tightly controlled and aligned with jurisdiction-specific regulations like GDPR. It provides interoperability between different systems and organizations, facilitating seamless data sharing and access control across diverse participants in data spaces.

Similarly, the use of SSI enhanced with ZKP offers improved privacy through selective disclosure and data minimization, ensuring that participants in the data space maintain control over their personal information. It complies with regulations like GDPR and promotes interoperability through the use of standards such as DIDs and VCs.

With the aim of enhancing the current mechanisms and the privacy and confidentiality of data spaces, several additional areas are being researched and explored. Definition of policies, both for sticky policies and policy enforcement mechanism, need to be optimized to offer a better experience to the user. Users need to be aware of what kind of policies they can enforce and how in a friendly way. Additionally, the integration with a consent/contract definition is being explored so that the policies outlined in the contract are enforced through the different mechanisms. In the scope of SSI, further alignment of the approach with initiatives like eIDAS2 [37], particularly improving upon them with advanced privacy properties. Another line is addressing the transition of the cryptographic operations used for authentication into post-quantum environments.

Advanced privacy-preserving enablers are being explored such as Confidential Computing by means of Trusted Execution Environments (TEEs), blockchain systems or additional anonymity support.

Additionally, the three different assets showcased in this paper are currently under further development and integration within the framework of data spaces as part of the two European projects TANGO [38] and TITAN [39], as mentioned before. TANGO project aims to develop a secure, trustworthy and environmentally sustainable data management ecosystem to provide a high impact solution within several domains like transport, e-commerce, finance, public administration or tourism supporting numerous beneficiaries across Europe. The core components of its architecture include a data government framework, distributed privacy-preserving data management and storage and a distributed trust management framework which are developed in alignment with IDSA, Gaia-X and FIWARE data spaces initiatives. Similarly, TITAN provides a platform solution for confidential, secure and privacy-preserving data processing and collaboration to enrich the EOSC Interoperability Framework. TITAN framework is to be demonstrated in public administration and healthcare sectors. Among its main innovations are confidential data processing based on Trusted Execution Environments (TEEs) and scalable and fine-grained decentralized blockchain-based access control, ensuring a truly end-to-end secure data lifecycle.

# References

1. Why data spaces? https://datos.gob.es/en/blog/why-data-spaces. Accessed 11 June 2024
2. LDAP. https://ldap.com/. Accessed 19 June 2024
3. What is OpenID Connect? https://openid.net/developers/how-connect-works/. Accessed 19 June 2024
4. An overview of limitations and approaches in identity management. https://doi.org/10.48550/arXiv.2301.00442. Accessed 30 Aug 2024
5. Integration of Self-Sovereign Identity into Conventional Software using Established IAM Protocols: A Survey. https://dl.gi.de/items/f64e3367-76e4-4216-ae8d-94281ae57150. Accessed 18 Aug 2024
6. Zero knowledge proofs of identity. https://doi.org/10.1145/28395.28419. Accessed 18 Aug 2024
7. Privacy by Design: The 7 Foundational Principles. https://student.cs.uwaterloo.ca/~cs492/papers/7foundationalprinciples_longer.pdf. Accessed 18 Aug 2024
8. Anonymization Techniques for Privacy Preserving Data Publishing: A Comprehensive Survey. https://doi.org/10.1109/ACCESS.2020.3045700. Accessed 18 Aug 2024
9. On Using Encryption Techniques to Enhance Sticky Policies Enforcement. https://research.utwente.nl/en/publications/on-using-encryption-techniques-to-enhance-sticky-policies-enforce. Accessed 18 Aug 2024
10. Challenges in the Emergence of Data Ecosystems. https://www.researchgate.net/profile/Joshua-Gelhaar/publication/341930759_Challenges_in_the_Emergence_of_Data_Ecosystems/links/5ed9edb8299bf1c67d41ad09/Challenges-in-the-Emergence-of-Data-Ecosystems.pdf. Accessed 18 Aug 2024
11. Reference Architecture IDSA. https://internationaldataspaces.org/offers/reference-architecture/. Accessed 18 June 2024
12. Industrial Data Space Architecture Implementation Using FIWARE. https://www.mdpi.com/1424-8220/18/7/2226. Accessed 18 June 2024
13. What Does it Take to Connect? Unveiling Characteristics of Data Space Connectors. https://scholarspace.manoa.hawaii.edu/server/api/core/bitstreams/1c6cb422-d15f-4ed4-a3fa-24890d8f635a/content. Accessed 18 June 2024
14. General Data Protection Regulation (GDPR). https://gdpr.eu. Accessed 18 June 2024

15. International Data Spaces. https://internationaldataspaces.org. Accessed 18 June 2024
16. Gaia-X: A Federated Secure Data Infrastructure. https://gaia-x.eu. Accessed 18 June 2024
17. FIWARE- Open APIs for Open Minds. https://www.fiware.org. Accessed 18 June 2024
18. European Open Science Club. https://eosc.eu. Accessed 18 June 2024
19. Data Spaces Support CentreAlthough the adoption of SSI is still in its early stages in enterprise environments, its ability to integrate with traditional systems through specific integration patterns offers a promising solution to overcome the limitations of centralized identity approaches. https://dssc.eu. Accessed 18 June 2024
20. The Data Spaces Business Alliance. https://data-spaces-business-alliance.eu/. Accessed 18 June 2024
21. Data Act. https://digital-strategy.ec.europa.eu/en/policies/data-act. Accessed 18 June 2024
22. DSSC Data Spaces Blueprint v1.0. https://dssc.eu/space/BVE/357073006/Data+Spaces+Blueprint+v1.0. Accessed 18 June 2024
23. DSBA Technical Convergence. https://data-spaces-business-alliance.eu/wp-content/uploads/dlm_uploads/Data-Spaces-Business-Alliance-Technical-Convergence-V2.pdf. Accessed 18 June 2024
24. World Wide Web Consortium (W3C). https://w3c-ccg.github.io. Accessed 18 June 2024
25. OpenID Foundation. https://openid.net. Accessed 18 June 2024
26. European Blockchain Services Infrastructure (EBSI). https://hub.ebsi.eu. Accessed 18 June 2024
27. Verifiable Credentials Data Model v2.0. https://www.w3.org/TR/vc-data-model-2.0/. Accessed 18 June 2024
28. Verifiable Credentials Data Model v2.0. https://www.w3.org/TR/did-core/. Accessed 18 June 2024
29. Verifiable Credentials Data Model v2.0. https://openid.net/specs/openid-connect-self-issued-v2-1_0.html. Accessed 18 June 2024
30. OpenID for Verifiable Presentations v1.0. https://openid.net/specs/openid-4-verifiable-presentations-1_0.html. Accessed 18 June 2024
31. OpenID for Verifiable Credential Issuance. https://openid.net/specs/openid-4-verifiable-credential-issuance-1_0.html. Accessed 18 June 2024
32. Implementation and evaluation of a privacy-preserving distributed ABC scheme based on multi-signatures. https://www.sciencedirect.com/science/article/pii/S2214212621001824. Accessed 18 June 2024
33. Pointcheval-Sanders multi signatures. https://eprint.iacr.org/2015/525. Accessed 18 June 2024
34. walt.id: Digital identity and wallet infrastructure. https://walt.id. Accessed 18 June 2024
35. IDS Reference Architecture Model v4.0. https://docs.internationaldataspaces.org/ids-knowledgebase/v/ids-ram-4. Accessed 18 June 2024
36. eXtensible Access Control Markup Language (XACML) Version 3.0. http://docs.oasis-open.org/xacml/3.0/xacml-3.0-core-spec-os-en.html. Accessed 18 June 2024
37. Electronic Identification, Authentication and Trust Services. https://digital-strategy.ec.europa.eu/en/policies/discover-eidas. Accessed 18 June 2024
38. Home | Tango Project. https://tango-project.eu. Accessed 18 June 2024. https://digital-strategy.ec.europa.eu/en/policies/discover-eidas
39. Titan EOSC Project. https://titan-eosc.eu. Accessed 18 June 2024

# The AURORAL Privacy Approach
# for Smart Communities Based on ODRL

Andrea Cimmino$^{(\boxtimes)}$ (ID), Juan Cano-Benito (ID), and Raúl García-Castro (ID)

Ontology Engineering Group, Universidad Politécnica de Madrid, Madrid, Spain
{andreajesus.cimmino,juan.cano,r.garcia}@upm.es

**Abstract.** Over the past decade, smart cities have developed significantly, while rural areas have lagged. The AURORAL project aims to
bridge this gap by offering a decentralised, semantically interoperable
platform tailored for rural concerns. This platform not only facilitates
the exchange of data but also adds features such as data validation or
privacy policies for the users. This paper explores how AURORAL uses
the ODRL language to allow users to establish and enforce privacy policies, thereby enhancing data privacy control.

**Keywords:** Data privacy · ODRL policies · Semantic interoperability

## 1 Introduction

In the last decade, the digitisation of environments to provide improved services,
digital goods, and potential business has focused on cities, leading to the creation
of smart ecosystems [1], that is, smart Cities. To this end, a wide range of
solutions have been presented to address common challenges such as semantic
interoperability, data decentralisation, privacy, and security [2].

However, since research has focused on improving and evolving smart cities,
other adjoining areas of cities have experienced slower or even no evolution.
In particular, one of the most affected areas is the rural [3]. The differences
between these areas have highlighted the need to evolve smart cities into smart
communities; making them more inclusive with adjoining areas and shifting from
smart cities to smart communities [4].

Despite the large research effort developed in the context of smart cities [5].
Most of these proposals fall short when contextualised in the smart communities
areas due to a wide number of factors, e.g., social, physical, or technological.
In particular, rural areas require two main features, such as semantic interoperability and privacy. For this reason, the European project AURORAL aims to
boost a novel ecosystem for smart communities to fill the gap between rural and
urban areas. To this end, it provides a fully decentralised semantic interoperable
platform to not only for exchanging data, but also for creating value on top of it.

This work is partially funded by the European Union's Horizon 2020 Research and Innovation Programme through the AURORAL project, Grant Agreement No. 101016854.

M. Presser et al. (Eds.): GIECS 2024, CCIS 2328, pp. 89–100, 2025.
https://doi.org/10.1007/978-3-031-78572-6_6

In this paper, the novel approach adopted in the AURORAL project to define privacy policies in the context of smart communities is described. The AURO-RAL platform allows practitioners to define and apply privacy policies based on the W3C standard ODRL language [6]. This approach is especially suitable, since it allows practitioners to define privacy conditions for accessing the data differently from the common approach based on white-lists or black-lists. These ODRL-based policies allow one to define conditions to access endpoints of top of data, which can be local or provided by a third party server. For instance, in a rural area, using the GPS data of an animal, more detailed information about it may become accessible only if its geographical point value is outside a fence, i.e., geofencing.

In order to showcase a real-world application of AURORAL the article presents one of its scenarios that is a smart laboratory at the Universidad Politécnica de Madrid, i.e., SmartLab. In this context, several devices are integrated through AURORAL to make their data available. However, due to privacy requirements, the scenario is configured with an ODRL-based privacy policy that grants access to the data only during a range of hours and revokes the access the rest of the time. The main benefit of using the AURORAL approach, as shown in this scenario, is that no coding is required to express such conditions.

The rest of this article is structured as follows. Section 2 presents a short overview of proposals focusing on smart cities for large IoT integration; Sect. 3 presents the AURORAL initiative, in particular, the semantic interoperability approach and the privacy approach; Sect. 4 provides an overview of the different domains in which AURORAL has been adopted and the details of how it has been adopted in a smart laboratory; finally, Sect. 5 recaps the findings and conclusions.

## 2   Related Work

Conventional IoT architectures are typically hierarchical and lack considera-tion of privacy and data sovereignty, resulting in challenges when it comes to sharing data between multiple parties. To enable effective data sharing, digital ecosystems must priorities interoperability and open standards. Several Euro-pean projects and initiatives have already addressed this need by integrating a wide range of sensors and developing services that leverage or enhance the information collected, namely: SymbIoTe [7], AGILE [8], bIoTope [9], INTER-IoT [10], or TagItSmart [11], or Big IoT [12], VICINITY [13].

AURORAL builds on the efforts of previous European projects and initia-tives, such as VICINITY, and includes novelties such as the decentralisation of metadata storage, a novel semantic interoperability approach, decentralized discovery, and its ODRL-based privacy approach [6]. AURORAL also aims to contribute to standardisation and interoperability in IoT, positioning itself as one of the first frameworks that provides trusted environments for sharing data in smart communities. In fact, AURORAL is mainly based on the W3C stan-dards.

## 3   AURORAL Platform

AURORAL is a platform for smart communities and rural areas that provides a flexible and interoperable ecosystem, facilitating the integration of data sources (vertical services) and allowing the creation of value-added services with the data available on the platform (horizontal services). The AURORAL platform consists of a set of distributed nodes deployed on the owners' premises that exchange data through the AURORAL middleware (an XMPP network); as shown in Fig. 1. However, the ownership of the data remains unchanged since no data is stored in the platform. The AURORAL architecture aims to facilitate the integration of different vertical tools and services for a variety of rural domains, as described in Fig. 2.

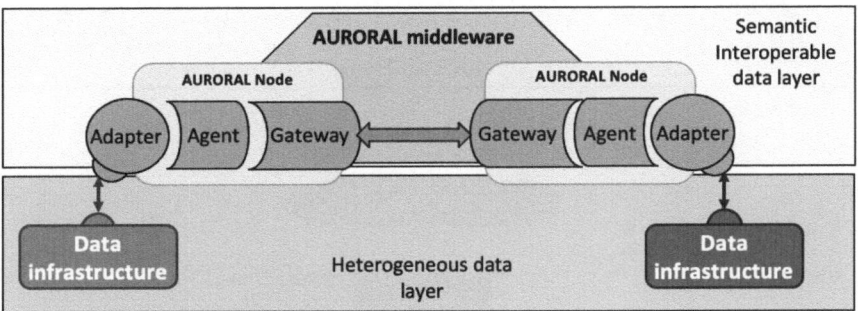

**Fig. 1.** AURORAL functionality.

The AURORAL nodes are one of the pillars on the AURORAL platform, Fig. 3 shows in detail the different components within. The nodes are configured with Thing Descriptions [14] in order to the platform to become aware of existing resources, and enable discovery among nodes. However, nodes are not allowed to discover others if their owners have not signed a contract that agrees on the exchange of data beforehand. Note that the AURORAL node is almost built on top of open standards from the W3C.

### 3.1   Semantic Interoperability

In order to achieve semantic interoperability in AURORAL, one of the main pillars is the AURORAL ontology[1]. This ontology consists of several modules that model specific subdomains of rural areas and smart communities. The ontology has been built following the LOT methodology [15].

All data being exchanged in the AURORAL middleware must adhere to the following semantic interoperability requirements: data must be in RDF format, preferably in JSON-LD 1.1 serialisation, and be correctly expressed following the

---

[1] https://auroral.iot.linkeddata.es/.

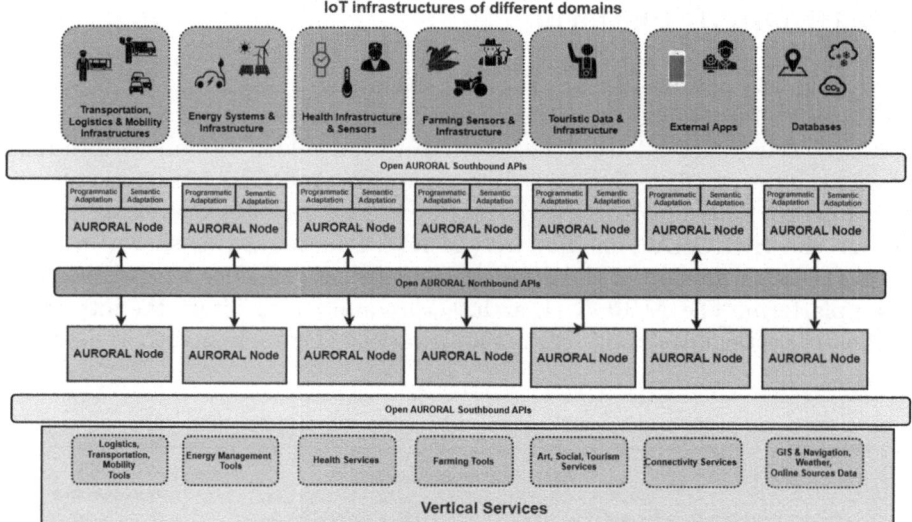

**Fig. 2.** AURORAL Architecture.

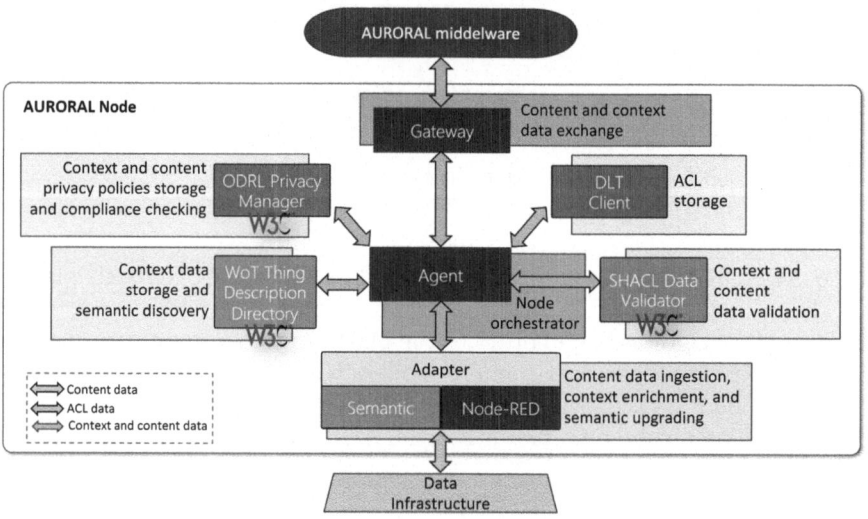

**Fig. 3.** AURORAL Node Architecture.

AURORAL ontology. In fulfilling these requirements, semantic interoperability is achieved thanks to the consensus on the format and model of the data. Only data that meet these requirements are allowed to be exchanged in the middleware. To ensure this, the AURORAL nodes count with the **Shapes Constraint Language (SHACL) Data Validator** that verifies the syntax of the data and ensures that it correctly follows the ontology model.

Rural areas and smart communities rely on a wide range of heterogeneous data sources that do not follow the AURORAL semantic interoperability requirements. In order to address this issue, the AURORAL node counts with the component **Adapter**[2] which has two implementations, the *Semantic Adapter* and the *RED-Node Adapter*. Both implementations aim at performing the same task, that is, adapting heterogeneous data so that they meet the semantic interoperability requirements. However, the way in which they achieve that goal is different as well as their capabilities. A third option is a *hard-coded adapter* that is developed from scratch and must implement the necessary interfaces to integrate this adapter with the AURORAL node.

The *Node-RED Adapter* is based on the Node-RED workflow framework[3]. This adapter was developed to simplify the integration with AURORAL since it only needed a user to define their data sources and the data values that must be integrated into AURORAL using a GUI. However, the main limitation of this adapter is that it sticks exclusively to data that can be expressed uniquely with the IoT ontology module[4]. As a result, data outside the IoT domain or with complex schemes cannot be integrated using this adapter.

The *Semantic Adapter* is based on the Helio framework [16]. This adapter takes as input a translation mapping that contains the pointers to the sources of data and a set of translation rules. Although this adapter supports any kind of data regardless of their domain or complexity in their schemes, it does not provide a GUI to build the mappings. As a result, it has a higher learning curve than *Node-RED Adapter*, but it also allows performing more complex translations regardless of the data domain.

The *hard-coded adapter* can be developed with any coding language as long as it meets two requirements. First, it has to deploy a REST service that implements the AURORAL interfaces, and second, the data published by the service must follow the semantic interoperability requirements. Although the benefits of this adapter are similar to those of *Semantic Adapter*, since it allows one to perform complex translations regardless of the data domain, it may even have a higher learning curve than the *Semantic Adapter*. These adapters are only suitable for practitioners experienced with coding and REST APIs; otherwise, they entail a high integration cost.

Finally, although practitioners may rely on any of the three adapters, the quality of the data produced by them is not guaranteed. It is worth mentioning that it is important to note that any communication through the middleware is first validated by the **SHACL Data Validator**. As a result, regardless of the adapter used in a node AURORAL ensures that the data exchanged are always semantic interoperable and fulfils the interoperability requirements established in the platform.

---

[2] https://auroral.docs.bavenir.eu/adapters/intro/.

[3] https://nodered.org/.

[4] https://auroral.iot.linkeddata.es/def/adapters/index.html.

### 3.2    Privacy Based on the Open Digital Rights Language

The AURORAL platform has two privacy mechanisms for data exchange. On the one hand, AURORAL nodes are not allowed to exchange data through the middleware if first the practitioners of two nodes do not sign an agreement (white-list access alike). On the other hand, once this agreement is settled, practitioners can specify privacy policies for their data which are evaluated on top of the white list approach. These privacy policies are one of the novelties that AURORAL provides.

The AURORAL nodes count with the **ODRL Privacy Manager** that stores semantic privacy policies expressed according to the Open Digital Rights Language (ODRL), which is a standard promoted by the W3C [6]. Although this language has a wider scope than just privacy, in AURORAL ODRL is used exclusively for defining data access privacy policies. Furthermore, the ODRL language is combined with RDF materialisation techniques, allowing practitioners to define privacy policies to access data based on data values, information sent in a request, or external services [17].

The **ODRL Privacy Manager** is set up by the practitioner who owns an AURORAL node. It provides a CRUD API to manage ODRL policies, which are JSON-LD 1.1 documents that follow the ODRL vocabulary [18]. In addition, this service validates the policies that have been registered to ensure their correctness.

Figure 4 shows how the AURORAL node handles data requests from other nodes through the middleware; in particular, how a remote request is successfully handled. As can be observed, any request is first evaluated against the **ODRL Privacy Manager** (steps 1 to 3). This component searches for possible ODRL policies related to the endpoint targeted by the request (step 4). In the event that an ODRL policy exists, the component evaluates the policy and, as a result, informs the agent if the access to the endpoint is granted or revoked (step 5). Based on the **ODRL Privacy Manager** output, the **Agent** returns the data related to the requested endpoint or an error message.

## 4    AURORAL Adoption

AURORAL project counts with a wide range of applications areas from different domains related to rural areas and smart communities (i.e., energy, mobility, tourism, health and agriculture). Table 1 shows the different domains, use cases, and pilots adopting the AURORAL solution. Furthermore, this section aims to show how the smart laboratory at the Universidad Politécnica de Madrid (UPM) has adopted the AURORAL technological stack. In particular, from the semantic interoperability and privacy point of view.

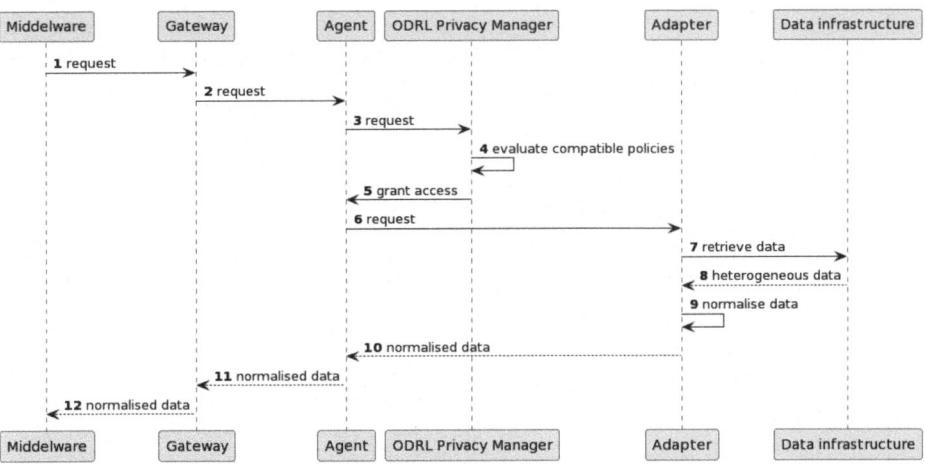

**Fig. 4.** Sample diagram sequence showing successful access to data infrastructure.

**Table 1.** Regions, domains and use cases of the AURORAL platform.

| Pilot region | Energy Use Case | Mobility Use Case | Tourism Use Case | Health Use Case | Farming Use Case |
|---|---|---|---|---|---|
| Alentejo, Portugal | 2 | 2 | 2 | 1 | 1 |
| Halogaland, Norway | 0 | 1 | 2 | 1 | 0 |
| Lapland, Finland | 0 | 3 | 0 | 0 | 0 |
| Penedès, Catalonia | 1 | 0 | 0 | 0 | 0 |
| Piedmont, Italy | 1 | 0 | 1 | 0 | 2 |
| Southern Burgenland, Austria | 1 | 2 | 1 | 0 | 0 |
| Vasterbotten, Sweden | 1 | 0 | 0 | 0 | 0 |

## 4.1 SmartLab at Universidad Politécnica de Madrid

The SmartLab is a laboratory at the Universidad Politécnica de Madrid that has
been sensorized with different devices that measure eight magnitudes, among
which temperature, occupancy, humidity, pressure, and sound. The SmartLab
aims to improve the workplace by making smart choices using sensor data or
interacting with researchers within the laboratory. For example, one of the goals
of the SmartLab is that when there is too much noise or when the amount of
CO2 is high, it prints a message as a notification on a screen.

In this context, an AURORAL node has been deployed in the SmartLab configured with the **Semantic Adapter**. In the laboratory, several devices push
their data into a Home Assistant, an open-source software for home automation,
designed to be an IoT ecosystem-independent integration platform and central
control system for smart home devices[5]. The **Semantic Adapter** relies on the
mapping described by Listing 1 to translate their data directly from the home
assistant into an equivalent payload that is accessible through the AURORAL

---

[5] http://smartlab.auroral.linkeddata.es.

node of the laboratory[6], which is expressed in JSON-LD 1.1 and follows the AURORAL ontology module for IoT that is based on Smart Applications REFerence (SAREF) [19].

```
1  <#assign jpath=handlers("JsonHandler")>
2  <#assign config="{ \"method\" : \"GET\", \"url\" : \"http://smartlab.auroral.linkeddata.
      es/api/history/period?filter_entity_id=sensor.temperature\", \"headers\":{\"
      Authorization\": \"Bearer Token\"}}">
3  <#assign homeassistant=providers("HttpProvider", config)>
4  {
5      "@context" : "https://auroralh2020.github.io/auroral-ontology-contexts/adapters/
          context.json",
6      "id" : "[=jpath.filter("S.[*].[0].entity_id",homeassistant)]",
7      "@type":"weatherSensor",
8      "measurement" : [
9          <\#list jpath.filter("S.[*].[*]",homeassistant) as fragment>
10         {
11         "timestamp":"[=jpath.filter("S.last_updated",fragment)]",
12         "value":[=jpath.filter("S.state",fragment)],
13         "type":"number",
14         "isMeasuredIn":"degree",
15         "property":"ambientTemperature"
16         }
17         <\#if fragment?is_last><\#else>,</\#if>
18         </\#list>
19     ]
20 }
```

Listing 1: Mapping for translating SmartLab device data for AURORAL

Furthermore, the AURORAL node of SmartLab is governed by an ODRL-based privacy policy described in the Listing 2. This policy allows or revokes the access to the data provided by the node depending on the time at which a request is performed, allowing the access only during hours which are foreseen that the researchers may stay at the laboratory (08:00–18:00) and revoking the access the rest of the time. As can be observed, this policy relies on the time of the system where the policy is evaluated and, as a consequence, considers the time zone of Madrid.

```
1  {
2  "@context": "http://www.w3.org/ns/odrl.jsonld",
3  "@type": "Set",
4  "uid": "http://example.com/policy:1010",
5  "permission": [{
6      "target": "https://helio-dev.data.linkeddata.es/api/upm-sensors/data",
7      "action": "read",
8      "constraint": [{
9          "leftOperand":  {
10             "@value": "[=.now?time?iso("Europe/Rome")?replace('\\+.*','', 'r')]",
11             "@type": "xsd:time"
12         },
13         "operator": "gt",
14         "rightOperand":  {
```

---

[6] https://helio-rest.auroral-factory.linkeddata.es/api/ODRL_Devices_mapping/data.

```
15              "@value": "08:00:00",
16              "@type": "xsd:time"
17              }
18          },{
19          "leftOperand":  {
20              "@value": "[=.now?time?iso("Europe/Rome")?replace('\\+.*','', 'r')]",
21              "@type": "xsd:time"
22              },
23          "operator": "lt",
24          "rightOperand":  {
25              "@value": "18:00:00",
26              "@type": "xsd:time"
27              }
28      } ]
29  }]
30  }
```

Listing 2: ODRL privacy policy to access data if user is in position

The data generated by the laboratory also includes GPS data from users. In this context, the laboratory data and user data generated are sensitive information, and in order to preserve sensitivity, users should only access lab data when they are in a workspace, and the lab node should only provide the location of users when they are working. The mapping described by Listing 3 to retrieve the GPS position from the smart lab into an equivalent payload that is accessible through the AURORAL node.

```
1  <#assign jpath=handlers("JsonHandler")>
2  <#assign config="{ \"method\" : \"GET\", \"url\" : \"http://smartlab.auroral-factory.
       linkeddata.es/api/history/period?filter_entity_id=device_tracker.cano\", \"headers
       \":{\"Authorization\": \"Bearer Token\"}}">
3  <#assign homeassistant=providers("HttpProvider", config)>
4  {
5      "@context" : "https://auroralh2020.github.io/auroral-ontology-contexts/adapters/
          context.json",
6      "id": "[=jpath.filter("S.[*].[0].attributes.friendly_name",homeassistant)] GPS
          Device",
7      "@type":"monitoringSensor",
8      "measurement": {
9          "timestamp" : "[=jpath.filter("S.[*].[0].last_updated",homeassistant)]"
10     },
11     "property":{
12         "gpsLatitude" :  "[=jpath.filter("S.[*].[0].attributes.latitude",homeassistant
              )]",
13         "gpsLongitude" : "[=jpath.filter("S.[*].[0].attributes.longitude",
              homeassistant)]",
14         "gpsAltitude" : "[=jpath.filter("S.[*].[0].attributes.altitude",homeassistant)
              ]",
15         "devPosition" : "[=jpath.filter("S.[*].[0].attributes.gps_accuracy",
              homeassistant)]"
16     }
17 }
```

Listing 3: Mapping for retrieve GPS position dynamically

This policy described in the Listing 4 allows or revokes the access to the data provided by the node depending on the geographical position at which a request is performed, allowing the access only if the user is in the working area. As can be observed, this policy relies on a polygon that defines the work area and the GPS position data obtained from the smart lab.

```
1   <#assign jpath=handlers("JsonHandler")>
2   <#assign config="{ \"method\" : \"GET\", \"url\" : \"http://smartlab.auroral-factory.
        linkeddata.es/api/history/period?filter_entity_id=device_tracker.personal\", \"
        headers\":{\"Authorization\": \"Bearer Token\"}}">
3   <#assign juanassistant=providers("HttpProvider", config)>
4   <#assign config="{ \"method\" : \"GET\", \"url\" : \"https://helio-rest.personal.auroral
        .linkeddata.es/api/JuanPosition/data\"}">
5   <#assign juanData=providers("HttpProvider", config)>
6   <#assign configNoAccess="{ \"method\" : \"GET\", \"url\" : \"https://helio-rest.personal
        .auroral.linkeddata.es/api/JuanNoAccess/data\"}">
7   <#assign juanDataNoAccess=providers("HttpProvider", configNoAccess)>
8   <#assign policy>
9   {
10    "@context": ["http://www.w3.org/ns/odrl.jsonld",
11      {"geof" : "http://www.opengis.net/def/function/geosparql/"}
12    ],
13    "@type": "Set",
14    "uid": "http://example.com/policy:1011",
15    "permission": [{
16      "target": "https://helio-rest.personal.auroral.linkeddata.es/api/JuanPosition/data",
17      "action": "read",
18      "constraint":{
19          "leftOperand":  { "@value": "POLYGON ((-3.476486 40.63532, -3.484898
                40.623725, -3.468761 40.624376, -3.461895 40.634017, -3.476486 40.63532))
                ", "@type": "http://www.opengis.net/ont/geosparql#wktLiteral" },
20          "operator": "geof:sfContains" ,
21          "rightOperand": { "@value": "POINT(([=jpath.filter("S.[*].[0].attributes.
                longitude",juanassistant)] [=jpath.filter("S.[*].[0].attributes.latitude
                ",juanassistant)]))", "@type": "http://www.opengis.net/ont/geosparql#
                wktLiteral" }
22      }
23    }]
24  }
25  </#assign>
26  <@action type = "ODRL" data=policy; result>
27  <#assign resultado="[=result]">
28  <#if resultado== ('{}')>[=juanDataNoAccess]<#else>[=juanData]</#if>
29  </@action>
```

Listing 4: ODRL privacy policy for SmartLab based on dynamic GPS position

Note that this approach for privacy allows to define conditions over the data to grant or revoke the access to AURORAL endpoints. One of the main benefits of it is the fact that privacy policies do not require to be implemented as code, just specified. In addition, the language used to specify these policies is a W3C standard.

# 5    Conclusion

The focus on creating smart ecosystems and solutions for smart cities has slowed the adoption and evolution of these smart ecosystems in adjoining areas, such as rural areas. As a result, the need to evolve smart cities into smart communities has become crucial, as they are more inclusive with these adjoining areas. The AURORAL project aims to bridge the gap between rural and urban areas by providing a fully decentralised semantic interoperable platform to create value over the data within the platform.

AURORAL provides a novel privacy approach that consists of policies based on the W3C standard ODRL. These policies allow one to define conditions for accessing data endpoints based on local or remote data. For example, with these policies, practitioners can define geofencing policies. The main benefit of the AURORAL approach to privacy is that it does not require coding, just specifying the policies. In addition, the language used is a W3C standard that allows one to define policies outside the withe-list or black-list approach for accessing data. In general, the privacy approach of the AURORAL project has the potential to address the unique challenges faced by rural areas.

In order to showcase the AUROARL privacy approach, the article presents how the SmartLab at the Universidad Polotécnica de Madrid has adopted the AURORAL solution. Furthermore, this laboratory counts with a privacy policy that allows accessing the data of several devices deployed in the laboratory only during a range of time, that is, from 08:00 to 18:00. The rest of the time, the device's data are private.

# References

1. Sánchez-Corcuera, R., et al.: Smart cities survey: technologies, application domains and challenges for the cities of the future. Int. J. Distrib. Sens. Netw. **15**(6), 1550147719853984 (2019)
2. Kubler, S., et al.: IoT platforms initiative. In: Digitising the Industry Internet of Things Connecting the Physical, Digital and Virtual Worlds, pp. 265–292. River Publishers (2022)
3. Cunha, C.R., Gomes, J.P., Fernandes, J., Morais, E.P.: Building smart rural regions: challenges and opportunities. In: Rocha, Á., Adeli, H., Reis, L.P., Costanzo, S., Orovic, I., Moreira, F. (eds.) WorldCIST 2020. AISC, vol. 1161, pp. 579–589. Springer, Cham (2020). https://doi.org/10.1007/978-3-030-45697-9_56
4. Azgomi, H.F., Jamshidi, M.: A brief survey on smart community and smart transportation. In: 2018 IEEE 30th International Conference on Tools with Artificial Intelligence (ICTAI), pp. 932–939 (2018)
5. Rohen, M.: IoT EU strategy, state of play and future perspectives. In: Next Generation Internet of Things–Distributed Intelligence at the Edge and Human-Machine Interactions, pp. 1–8. River Publishers (2022)
6. Iannella, R., Villata, S.: ODRL Information Model 2.2 (2018)
7. Soursos, S., Podnar Žarko, I., Zwickl, P., Gojmerac, I., Bianchi, G., Carrozzo, G.: Towards the cross-domain interoperability of IoT platforms. In: 2016 European Conference on Networks and Communications (EuCNC), pp. 398–402. IEEE (2016)

8. Felfernig, A., Erdeniz, S.P., Azzoni, P., Jeran, M., Akcay, A., Doukas, C.: Towards configuration technologies for IoT gateways, vol. 73 (2016)
9. Werthmann, D., Hellbach, R.: Evaluation Report of the bIoTope Pilots (2017)
10. Ganzha, M., Paprzycki, M., Pawłowski, W., Szmeja, P., Wasielewska, K.: Semantic interoperability in the Internet of Things: an overview from the INTER-IoT perspective. J. Netw. Comput. Appl. **81**, 111–124 (2017)
11. Georgoulas, S., Krco, S., van Kranenburg, R.: TagItSmart–SmartTags for unlocking business potential. IEEE IoT Newslett. (2017)
12. Bröring, A., et al.: Enabling IoT ecosystems through platform interoperability. IEEE Softw. **34**(1), 54–61 (2017)
13. Cimmino, A., et al.: VICINITY: IoT semantic interoperability based on the web of things. In: 2019 15th International Conference on Distributed Computing in Sensor Systems (DCOSS), pp. 241–247. IEEE (2019)
14. Kaebisch, S., McCool, M., Korkan, E.: Web of Things (WoT) Thing Description 1.1 (2023)
15. Poveda-Villalón, M., Fernández-Izquierdo, A., Fernández-López, M., García-Castro, R.: LOT: an industrial oriented ontology engineering framework. Eng. Appl. Artif. Intell. **111**, 104755 (2022)
16. Cimmino, A., García-Castro, R.: Helio: a framework for implementing the life cycle of knowledge graphs. Semant. Web 1–27 (2022)
17. Cimmino, A., Cano-Benito, J., García-Castro, R.: Practical challenges of ODRL and potential courses of action (2023)
18. Iannella, R., Steidl, M., Myles, S., RodrÃguez-Doncel, V.: ODRL Vocabulary & Expression 2.2 (2018)
19. Daniele, L.M., Punter, M., Brewster, C., García Castro, R., Poveda, M., Fernández, A.: A SAREF extension for semantic interoperability in the industry and manufacturing domain. In: Enterprise Interoperability: Smart Services and Business Impact of Enterprise Interoperability, pp. 201–207 (2018). https://doi.org/10.1002/9781119564034.ch25

# Decentralized Management of IoT Platform Federations and Data Marketplaces

Ilia Pietri[1(✉)], George Darzanos[2], Georgios Spanos[3], Dimitra Karadimou[2],
Thanasis G. Papaioannou[2], George D. Stamoulis[2], Konstantinos Votis[3],
and Dimitrios Tzovaras[3]

[1] Intracom Single Member S.A. Telecom, Paiania, Greece
ilpiet@intracom-telecom.com
[2] Athens University of Economics and Business, Athens, Greece
[3] Information Technologies Institute/Centre for Research and Technologies Hellas,
Thessaloniki, Greece

**Abstract.** The Internet of Things (IoT) paradigm has been rapidly adopted in a plethora of application domains, promoting the development of a variety of smart services relying on data from different sources. As a consequence, Sensing-as-a-Service is also gaining popularity, enabling IoT data providers and consumers to share, exchange and trade IoT data. In such a data-driven ecosystem, data marketplaces play a pivotal role, which besides the need for interoperable solutions, necessitates also sophisticated market mechanisms that ensure trusted and secure data exchange. In this paper, we present the IoTFeds solution which offers a complete open source interoperability framework combined with blockchain technologies, and thus provides a decentralized federation management and marketplace platform. The IoTFeds platform enables trustworthy and secure transactions allowing the creation and monetization of composite IoT data services offered by multiple federated data providers.

**Keywords:** federated marketplaces · IoT data market · blockchain

## 1 Introduction

Data markets have recently experienced significant growth with the abundance of data in both scope and volume, and particularly in the Internet of Things (IoT) field. The numerous capabilities of exploiting such data collections through the use of data processing and analysis techniques and the continuous expansion of the application domains they cover has further motivated the ever-evolving IoT landscape.

As a result of the proliferation and diversity of the IoT applications, platforms, software and protocols, the IoT market is highly fragmented and the

© The Author(s) 2025
M. Presser et al. (Eds.): GIECS 2024, CCIS 2328, pp. 101–116, 2025.
https://doi.org/10.1007/978-3-031-78572-6_7

complexity of creating services exploiting the multitude of platforms and their connected IoT devices is even more challenging, while interoperability and security aspects are of utmost importance. Indeed, considering the huge number of connected IoT devices and the exponential growth of this market[1] that could be easily observed in Fig. 1, it is imperative to pay attention in the two aforementioned aspects. Therefore, with the emergence of solutions offering interoperability in the diverse IoT environments, covering the first critical aspect in the IoT paradigm, the interactions between IoT platforms, devices and applications involving the exchange of IoT data necessitate the second one, which is to have control over their access as well as the sharing data. Hence, the implementation of mechanisms for data markets that do not only enable the secure exchange of IoT data but also provide the necessary charging, invoicing and billing of transactions becomes essential.

The IoTFeds solution[2,3] introduces an innovative platform for the secure exchange of IoT data between IoT organizations (platform providers and consumers) developing an open source middleware that implements a fully distributed federation management and marketplace framework. The proposed solution is based on the outcomes of the H2020 symbIoTe project[4] combined with Distributed Ledger Technologies (DLT) such as Blockchain [12]. The symbIoTe software implements an IoT orchestration middleware that supports semantic, syntactic and organizational interoperability enabling IoT platforms federation and collaboration. The incorporation of the Blockchain technology into the IoTFeds ecosystem enables the creation of a decentralized environment bringing several benefits such as: the elimination of the need for a trusted third-party entity [20], novel and efficient business processes [9], transparency,

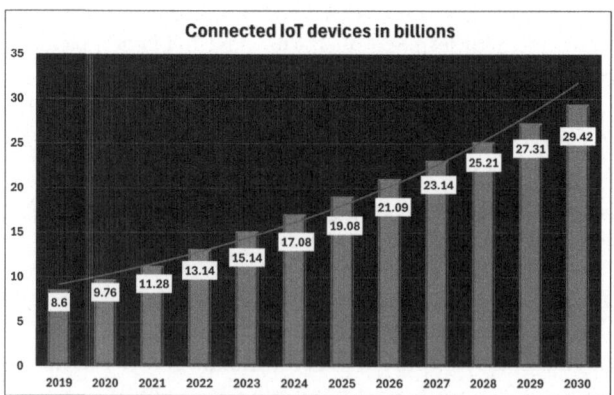

**Fig. 1.** IoT market growth.

and democracy [18], and finally, traceability and enhanced security [10]. However, blockchain integration may also bring some drawbacks that should not be neglected with the most important to be the system performance with respect to the transactions [18]. Associated challenges need to be considered when designing a blockchain solution for a more efficient development.

The IoTFeds platform provides all the necessary mechanisms to IoT organizations to form federations and jointly operate controlled markets for the secure and trusted data exchange and trading. The platform enables the creation of *federated (decentralized)* marketplaces where data services that may consist of data coming from different IoT providers can be built and exposed. It also offers a *global* marketplace where data services from any federated market can be promoted to make them available outside of the federation, including also organizations that do not necessarily belong to any federation. As opposed to related work for data-driven collaboration such as data space and marketplace frameworks [1,2,17], IoTFeds enables providers to collaborate and combine their data towards the composition of new data services, encapsulating value sharing and compensation mechanisms.

The decentralized IoT data market is implemented based on the following mechanisms:

- *Decentralized* IoT federation governance using *smart contracts* and voting procedures.
- Semantic *interoperability* within federation and globally.
- Secure and controlled access to IoT data based on *ABAC* mechanisms and *blockchain*.
- Discovery and *distributed search* of IoT data services.
- *Reputation* and *trust* methods.
- *Market mechanisms* with (non-)monetary considerations.

The purpose of this article is the general presentation of the main components composing IoTFeds platform avoiding the technical analysis of each component. The rest of the paper is organized as follows. Section 2 presents the main concepts in the IoTFeds ecosystem. Section 3 presents the platform's architecture with its main architectural subsystems, while Sect. 4 concludes the article.

## 2   IoTFeds Ecosystem

The IoTFeds platform enables IoT providers to establish federations and jointly provide data services to other IoTFeds members. It can support two types of data services; *single-source data services* offer direct access to data retrieved from a single data source (resource) e.g., an IoT device or data storage medium, and *data products* are bundles of multiple data sources that may come from several providers. Within the IoTFeds ecosystem different federation types may exist.

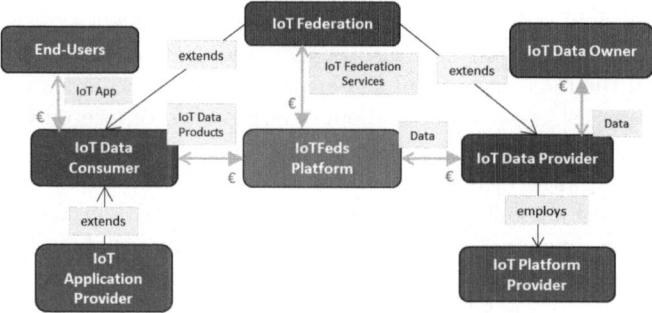

**Fig. 2.** IoTFeds value network.

## 2.1  Value Network

The value network presented in Fig. 2 illustrates the actor roles existing in the federated IoT data markets ecosystems (including IoTFeds) and their business relationships. The IoTFeds platform role positioned at the center of the ecosystem is a role adopted by all the entities contributing to the decentralized realization of IoTFeds platform, which involves the federation interoperability framework *symbIoTe* and its extensions to support the federated markets as well as the *blockchain* network for supporting the decentralized operation of IoTFeds platform and federations formed therein. The IoT data *provider* and *consumer* roles are highly prominent in the IoTFeds ecosystem, being the entities interacting with the IoTFeds platform to form federations and contribute to the formation of collective data services (data provider role) or consume data services (data consumer role).

Each federation is formed by a set of IoT data providers and potentially consumers, however they appear as a separate role in the value network because the IoTFeds platform treats federations as separate *Decentralized Autonomous Organizations* (DAOs) [19], offering the appropriate decentralized mechanisms for their democratized management and governance. An IoT data consumer can also adopt the IoT application provider role utilizing the data obtained by the IoTFeds marketplaces for applications (e.g. a smart parking) offered to end-users (e.g. drivers). Finally, an IoT data provider may own an IoT platform managing data collected from multiple IoT data owners who have ownership of the data coming from IoT devices they maintain.

## 2.2  Federation Types

Based on its members roles, federations can be categorized to *data providers'* or *mixed federations*. In a data providers' federation *all* federation members *must* contribute IoT data to the federated market, having the role of the IoT data provider. At the same time all members are able to consume IoT data contributed by other federation members, adopting the role of data consumer. In a *mixed federation* the members can adopt either or both roles. The key difference is that a member does not necessarily have the provider role but may only have the consumer role. Based on the market policies adopted by a federation, the federations can further be categorized to *closed* and *hybrid* federations. In a closed federation the federated data services can be only consumed by the federation members through a *federated marketplace*. In a hybrid federation, the federated data services can be exposed and made available to non-members through the open *global marketplace* of IoTFeds.

## 2.3   Marketplaces

All data consumers have access to the IoTFeds global marketplace, while access to a federated market is allowed to the members of the respective federation. Products can only be composed within federated marketplaces from data sources (resources) available within its federation. The products created within the federated marketplaces can also be promoted to the global marketplace as prefabricated products. Composing new data services through packaging is not allowed in the global marketplace. Resources and products within federation and marketplaces are discovered and accessed, filtering them based on user's selected criteria (e.g., trust, price etc.) and enforcing appropriate access control policies.

At the market level, the IoTFeds ecosystem has its own currency, named *FedCoin*, through which market transactions are processed, while data access is controlled through the use of access tokens. Markets can be distinguished to product and data source (resource) markets based on their data services type. In markets using the data product concept (product markets) access to data sources is only enabled through the definition of the product, i.e., a data source can itself be a product to be made available for accessing. Making data sources available for accessing directly makes sense only for data source markets, which do not consider the data products concept. In product markets, access to a product's data is controlled through the use of *access tokens* issued by the IoTFeds *Blockchain as a Service* (Baas) subsystem upon the purchase of the product from the respective marketplace (global or federated). In markets where transactions are made at the data source (resource) level, tokens that provide direct access to data sources are managed and traded through the symbIoTe subsystem. In both market types, i.e., the exchange of data sources or products, may be supported offering the potential to data consumers to place their unused access tokens for exchange with other consumers' tokens which provide access to data sources or products they have interest on. Both types of tokens, i.e., resource and product access tokens, are further described in Sect. 3.3.

*IoTFeds Example Instance.* Figure 3 shows an instance of the IoTFeds ecosystem with three data providers who are also consumers, and two data consumers-only. Providers 1 and 2 form federation 1 (federation of providers), while providers 2 and 3, as well as consumer B form federation 2 (mixed federation). As shown, provider 2 participates in both federations. The providers register all or a subset of their resources to the federated marketplace of each federation they participate. Each of the members of the federation can create

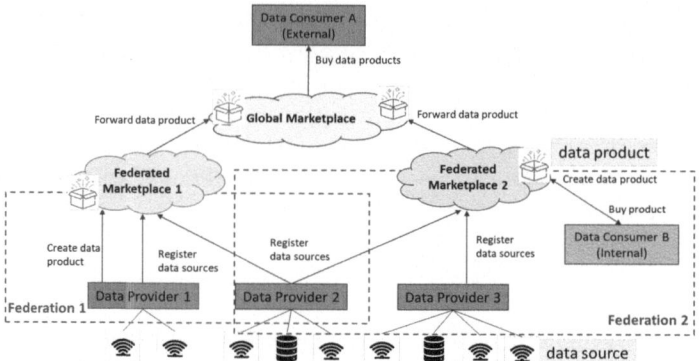

**Fig. 3.** IoTFeds ecosystem – with the different markets and interactions.

federated data products by combining any resources available in the marketplace. In Fig. 3, provider 1 creates a product in federated marketplace 1, while consumer B creates a product in federated marketplace 2. The created products can be bought and consumed by their creators or any other federation members. Both federations are open, meaning that the created products are forwarded to the IoTFeds global marketplace and are available to non-members of the federation. In the aforementioned example, data consumer A can buy products exposed by the federated marketplaces. A different pricing policy can be applied to data products forwarded to the global IoTFeds marketplace (compared to the federated marketplace) so that prices are more favorable for consumers within the federation. For marketplaces that do not adopt the data product concept, the direct accessing of data sources is performed only through the federated markets.

## 3    IoTFeds Architectural Elements

The core subsystems of the IoTFeds platform that work collaboratively to provide the required functionalities in the IoTFeds ecosystem include the symbIoTe middleware, the BaaS software and the Marketplace component. The creation, operation and management of the federations as well as their federated marketplaces are supported by the subsystems of symbIoTe and BaaS, with the marketplace acting as the online transactional place that exposes the IoTFeds market to its users. The symbIoTe software allows IoT platforms to expose selected IoT devices and their data to third-party systems as well as to interact with each other or collaborate to create horizontal solutions through its semantic interoperability approach and a unified and secure interface to access IoT resources. On the other hand, the BaaS subsystem enables the management of user identities and IoT platforms to ensure security and allow tracking of IoTFeds platform data, integrating distributed mechanisms for pricing each transaction and calculating the trust and reputation levels of data providers and their resources. Through the marketplace, IoT providers can collectively trade IoT data that

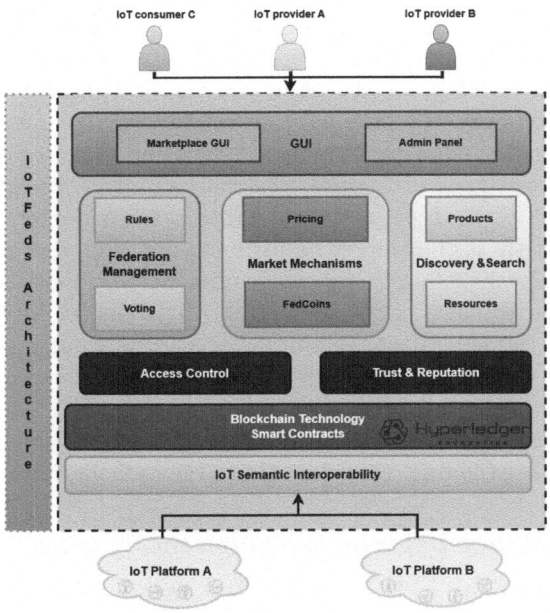

**Fig. 4.** IoTFeds architecture.

may come from several data sources to potential IoT consumers that wish to buy and access them.

As shown in Fig. 4, the decentralized IoT data market is operated by the following architectural parts of the IoTFeds platform: the IoT federation management, access control services (or secure access), distributed search and discovery services, trust and reputation system and the marketplace (or market mechanisms), discussed in the next section. The five architectural parts are provided by integrating the services and functionalities of its core subsystems, i.e., the symbIoTe middleware, the BaaS software and the Marketplace component. Blockchain technologies with the incorporated smart contracts and IoT semantic interoperability with the defined information models comprise the underlying level in the architecture. The IoTFeds services are exposed through high level APIs with an administration panel and a marketplace portal facilitating the management of the IoT federations and their associated marketplaces.

### 3.1   Democratized IoT Federation Governance

The implementation of the *federation management* mechanism in IoTFeds follows its own DAO approach [3] to achieve decentralized and autonomous federation governance with the creation of a relative smart contract. In the DAO approach any decision is made using voting conducted on the blockchain through a smart contract. Any member in the federation can potentially submit a proposal for consideration by the rest of the members. The request triggers a voting

procedure for the approval or rejection of the proposal. A member can participate in a voting procedure if they own governance tokens. The percentage of
the positive votes required to accept a proposal is specified in the *federation
rules*. The IoTFeds DAO solution is based on two smart contracts, one related
to the federation management and involved rules and actions (elaborated below)
and one related the voting procedure. The rules of an IoTFeds federation define
aspects such as: (*i*) the type of the federation (provider or mixed), (*ii*) the
type of its market (hybrid or closed) and the adopted market mechanisms, (*iii*)
the subscription details (if any), (*iv*) the supported verticals, (*v*) the supported
ontologies, (*vi*) governance rules supporting proposals for inviting/removing a
member to/from the federation, joining a federation, changing federation rules
etc., (*vii*) voting procedures rules offering a static template to parameterize in
each DAO federation, (*viii*) quality assurance rules specifying the quality met-
rics to be taken into account, the acceptable quality level and the management
mechanisms in case of failure, (*ix*) the coin to be used (fiat money or FedCoins).

## 3.2   Interoperability Between the Different Systems

Semantic interoperability in IoTFeds is based on the existing symbIoTe's solution that enables both interoperability by standardization (out-of-the-box) and
interoperability by mapping [21]. It relies on symbIoTe's *Core Information Model*
(CIM) with Extensions approach [7] with the definition of a basic information
model (ontology) that describes the basic entities and their information (vocabulary) that platform providers should use to describe their IoT devices and register them to the system. In symbIoTe, information models have been designed
to be generalised and abstract on the one hand but also detailed to enable flexibility to platforms adopting them while covering their needs [8]. The models
are inspired by commonly used ontology standards like SSN and SOSA[5]. The
approach used also allows platform providers to optionally register their own
platform specific information models (PIMs) to the system to include additional
device-specific details that cover their own needs. Based on semantic mapping,
additional device properties can be interpreted by other platform providers while
they can define and store to the system a correlation between their vocabularies
(mapping) on how data expressed using one ontology can be translated into the
terms of another ontology. A special category of PIM, the Best-practice Information Model (BIM) is also incorporated in the system to cover specific use
case domains for out-of-the-box interoperability between the platforms. Within
a federation, BIM can be selected to be used or a new PIM that extends CIM
can be agreed between the federation members and registered to facilitate data
and metadata transactions within it. Within the context of IoTFeds, BIM has
been extended to cover Smart City domain needs investigating state-of-the-art
ontologies like SAREF4CITY[6] and NGSI-LD[7] for smart city[8].

---

[5] https://www.w3.org/TR/vocab-ssn/.
[6] https://saref.etsi.org/saref4city/.
[7] https://www.fiware.org/smart-data-models/.
[8] http://www.symbiote-h2020.eu/ontology/smartcity_v1.

### 3.3    Access Control to IoT Data

IoT and blockchain technologies can be combined to control the conditions under which a user can access IoT devices and their data (e.g. when and for how long) creating a verifiable, decentralized and trustworthy environment on the one hand, while enabling the secure and reliable interconnection of a large number of devices on a network on the other hand. In IoTFeds, *Attribute-Based Access Control* (ABAC) mechanisms [14] to achieve secure data access and management at both resource and product levels are used (i.e. for its *access control services*). For instance, only the data provider has the privilege to create a federation and only the members of a federation have access to the federated marketplace.

*Resource-Level.* In the ABAC methodology, access to resources is controlled through the use of appropriate attributes (properties, roles or permissions associated with a resource in the system) implementing access control policies that make use of them to determine whether an action to a resource is allowed. The access policy defines a specific combination of attributes required to grant access to the IoT resource and is assigned to it by its owner. Therefore, a data consumer can access a resource only if it has a set of attributes that match the provider's predefined access policy.

*Product-Level.* ABAC mechanisms allow users to search, filter and access IoT resources and products in the participated federation and market according to their permissions. Authorization utilizes symbIoTe's ABAC mechanism combined with a security mechanism implemented at the blockchain network (BaaS subsystem). ABAC mechanism in symbIoTe allows to control user's access to IoT resources (data source level) defining a set of *policies* that cover the required cases, while user actions at product level are controlled by BaaS with the use of access tokens. Each user can be considered to have as a characteristic the specific tokens corresponding to the products in their possession, creating a *wallet* for each user at the *ledger* where their access tokens are stored. When a user successfully purchases a product, the user obtains an access token with the respective parameters (validity period, number of uses, frequency of access etc.). When a user requests access to a product's data, BaaS executes the respective search to the ledger for the user's wallet to locate the token with the name of the product and check its parameters to determine whether user's request can be served (e.g., for a multiple data access product, the remaining uses and if the date is inside the validity period are checked). Access control is done using smart contracts while the user is given the possibility through BaaS to control their wallet and the available tokens by retrieving its content from the ledger.

### 3.4    Trust and Reputation

In an IoT environment where diverse entities like IoT platforms, devices, services, applications and users interact at a high degree, establishing trust to ensure reliable transactions between trading entities can be a complex task. Reputation and trust systems are valuable tools for assessing the quality of the services

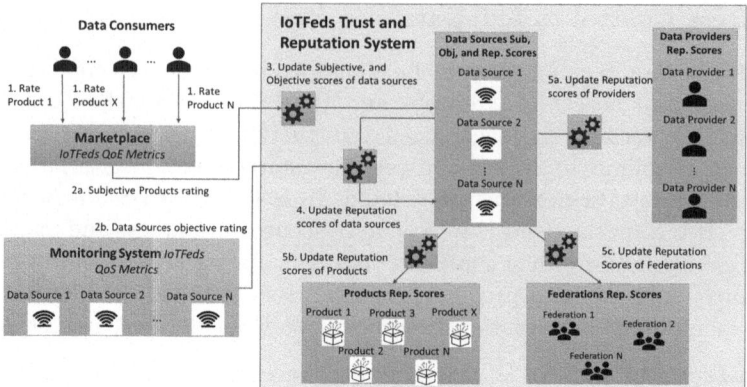

**Fig. 5.** Trust and reputation system.

provided, when the relevant information is incomplete, asymmetric or "hidden", while providing incentives to the entities to behave as expected and not to abuse the system. In the IoTFeds platform, the concept of reputation on multiple levels is introduced and leveraged as depicted in Fig. 5 and elaborated below.

– *Data sources.* The reputation score of a data source can facilitate the selection of appropriate quality data sources in product creation at a federated marketplace.
– *Data Products.* A product's reputation score is determined by the collective reputation score of its constituent data sources. In IoTFeds, the weighted sum aggregation approach [4,13] is adopted for the calculation. Product reputation scores may be leveraged by consumers for selecting appropriate quality pre-existing products in the federated or global marketplaces of IoTFeds.
– *Data Providers.* A data provider's reputation score is determined by the collective reputation score of its contributed data sources again utilizing the weighted sum aggregation method. Providers reputation scores are useful for adding/removing a member to/from a federation when its individual reputation score is higher/lower than the minimum reputation score set by the federation. Also, providers with higher reputation scores may have more influence on the decision making of the federation DAO.
– *Federations.* Similarly, the reputation score of a federation is determined by the collective reputation score of all data sources registered to the federation by all federation members utilizing the weighted sum aggregation method. Federation's reputation can be useful for an organization to select a trusted federation to join.

As described, the reputation scores of data sources are utilized for estimating the reputation scores in all other layers. However, certain inputs are needed to calculate each data source reputation. Two types of inputs are considered in IoTFeds *trust and reputation* system: (*i*) the *subjective rating* of products purchased by the consumers in federated or global marketplaces for a set of predefined Quality of Experience (QoE) parameters ("Easy of use", "Value for money", "Business Enablement", "Data completeness" etc.) and (*ii*) the *objective rating* of each data source based on performance measurements obtained from the monitoring system of the platforms on a set of predefined Quality of Service (QoS) parameters (service availability, response time, data precision etc.).

Each federation defines a QoE profile that identifies the set of parameters that are taken into account in product rating and their importance. Accordingly, a QoS profile is defined by the federation including target values for the QoS parameters. Based on the objective and subjective ratings, the IoTFeds Trust and Reputation system calculates and maintains an objective and a subjective score per data source which are combined to periodically update the reputation scores of data sources, products, providers and federations. All calculations (subjective, objective and reputation scores) are performed utilizing the appropriate smart contracts and the scores are maintained on-chain (BaaS subsystem). On the other hand, the data source QoS ratings (monitoring reports) and the product ratings are maintained off-chain (symbIoTe subsystem).

IoTfeds ensures the robustness of its reputation system against malicious users, such as data consumers who aim to damage a competitor's reputation or providers who report inaccurate performance estimates. Consumer attacks are mitigated by allowing only one objective rating per valid transaction, with an additional fee for each transaction. Provider false reporting is avoided through cross-checking with subjective ratings. Providers who consistently overestimate their performance lose trustworthiness, and their objective reports are excluded from reputation calculations.

## 3.5   Discovery and Distributed Search of Data Services

As mentioned earlier, data services in IoTFeds can be either *simple IoT resources* from a single IoT provider or more *complex products* (bundles of federated resources) promoted in the federated or global marketplace for purchase. Searching among the available data services is important for end users to select valuable resources or products meeting their needs based on selected criteria, such as location, data type, price and reputation. The symbIoTe platform offers a search mechanism for resources covering criteria like platform id, federation name, resource id, location and observed properties, while considering the filtering policies of the registered resources. Within IoTFeds, searching of resources and products covering criteria related to pricing and reputation is supported by BaaS. Thus, search in IoTFeds is realized on two levels. In the first stage, symbIoTe searches among the entire set of resources within the participated federation for the criteria it supports resulting in a filtered list of resources. At the second level, this list is further filtered and sorted in BaaS based on their price and/or reputation if necessary, while checking their validity. For products discovery, the search service of the marketplace is used supporting various criteria similar to the product's resources and combined with search in BaaS to additionally filter and sort them based on their price and trust values.

For both procedures, namely the discovery of resources and products, the last step before visualizing the results to the user is the *ranking* procedure [16]. It is indispensable for the proposed ranking algorithm to consider the reputation of resources and products by advertising (ranking higher in the results' list) the highly reputed ones. Thus, there is a close collaboration between trust & reputation system and the data services discovery system. More specifically, the ordering of the results depends on the products' and resources' reputation scores received from the ranking algorithm. However, a randomization factor is also considered in the proposed ranking algorithm to avoid a deterministic approach (i.e. having the same results with the same criteria) and offer a chance for a higher position in the results' list for new products having lower reputation scores. The aforementioned adopted ranking is met in the literature with the term stochastic ranking [6]. The whole procedure of the data services *search and discovery* is depicted in Fig. 6.

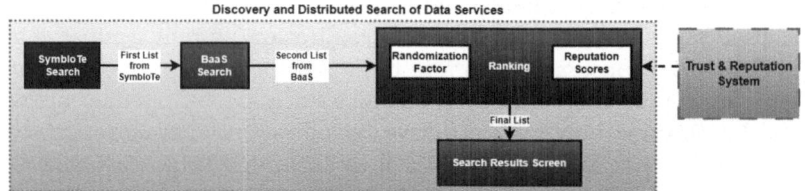

**Fig. 6.** IoTFeds Data Services Search.

### 3.6   IoTFeds Currency and Market Solutions

The IoTFeds ecosystem has its own currency, *FedCoin*, that is based on *ERC-20* fungible tokens [5]. FedCoins are created by the IoTFeds platform and are acquired by the users of the system instead of fiat money (euro, dollars, etc.). All the transactions in the marketplaces are performed in FedCoins. FedCoins are also used for micro-payments, namely transactions related to management aspects within the IoTFeds platform such as the federation governance and marketplace actions,e.g. product creation.

When a provider registers a data source in a federated marketplace, they declare its "value" in terms of FedCoins. This value expresses the cost of including this data source into a product capturing the cost for a single, most recent observation. Accordingly, each product is characterised by a price expressed in FedCoins, which must be paid by a buyer to acquire it. The product's price is based on the "values" of its constituent data sources and is expressed as a function of the (*i*) number of observations acquired (upon an observation request all data sources in the product are accessed), (*ii*) time window is which the acquired observations can be consumed and (*iii*) frequency of use. Data providers are then compensated according to the "value" and reliability of their data sources contributing to the purchased products.

There are multiple regimes of integrated IoTFeds *market* solutions. Each federated or global marketplace in IoTFeds may focus on different data services (data source or product services), and adopt different service pricing policies (static or dynamic), consumer charging mechanisms (monetary or nonmonetary) and provider compensation mechanisms. Next, three applicable in IoTFeds market solutions are highlighted.

*Data Product Market.* Data products are charged based on their use. Any member of the federation can search for data sources available within it to build the product of interest by paying a product creation fee to the IoTFeds platform. Providers declare the compensation they wish to receive for including the data source in a product as a function of observations number, consumption timewindow and frequency of use. These parameters are set by the buyer during the product's purchase. Each provider receives compensation for their contribution to the product depending on the data sources comprising the product and their declared compensation functions. Access to the product is controlled through the use of product access tokens (Sect. 3.3) managed in BaaS that specify the time period of accessing it and the observations frequency and number.

*Data Product Exchange.* This non-monetary exchange market solution is complementary to the above. In this type of "side" market, the consumers can exchange products they have purchased but have not totally consumed by placing to the exchange market. Each product placed for exchange and its details (expiration date, remaining number of uses, frequency etc.) is visible to all members of the market and any market member can submit a proposal by offering one of its purchased products in return. The concerned consumer is informed for the proposal to accept or reject it. The estimated current products' values can be displayed to the users to facilitate fairness in exchanges. If the offer is accepted the respective access tokens are transferred to their new owners' wallets.

*Data Source Exchange.* This is an alternative exchange mechanism in the data source level which is suitable for market federated solutions that do not adopt the data product concept and is applied to closed federations. In this solution, bartering takes place without any monetary implications with participants exchanging their respective data sources directly for access to data sources of interest. This approach is based on symbIoTe's *bartering & trading* mechanism where the concept of *Vouchers* [21] is introduced to allow mutual exchange of data. A data provider that wants to access a resource of another IoT platform (data provider) issues a voucher that grants the holder access to one of its resources and offers it to the other party to be consumed at a later date. In an IoTFeds market solution, when a provider requests access to another provider's data sources for a certain number of uses or duration, it offers in return a resource access token managed in symbIoTe promising access to its own platform's data sources for a corresponding number of uses or duration.

### 3.7   Intuitive Graphical User Interface

The developed platform includes an intuitive graphical user interface (GUI) to control and operate the decentralized market, as shown in the figure below, facilitating user interaction with the IoTFeds ecosystem and its services. The GUI consists of an administration panel and a marketplace portal that enable the management, configuration and operation of the IoTFeds federations and their associated marketplaces respectively. The administration panel is based on the extension and modification of symbIoTe's web-based GUI, providing a control panel for administration and configuration actions regarding the management and parameterization of the system's users, platforms and federations. Within IoTFeds, the administration panel is also accompanied with a marketplace portal operating as an online IoT data marketplace where IoT data providers and/or consumers can trade, buy and exchange IoT data services.

## 4   Discussion

This article described a decentralized federation management and marketplace framework presenting the architecture of the developed IoTFeds platform. The proposed solution enhances an open source IoT federation interoperability middleware with federated markets and blockchain network to enable the creation of IoT federations and their markets operation supporting the build of data product services from multiple federated resources. The platform follows a micro-services architecture deployed in a containerized environment for improved scalability and performance.

In the current approach data exchanges and interactions within federations are considered supporting the creation of products from data sources of the same federation. Although a data source may be shared in multiple federations participating in the creation of multiple products, each product can only be accessed through its federated market and optionally from the open global data market. Data trading mechanisms between federations and their marketplaces for cross-domain data products towards a cross-domain market operation could be studied. Also, multiple other alternative solutions may appear to IoTFeds based on the preferences of each federation, such as applying more sophisticated market mechanisms for optimal automated creation of products based on consumer requirements or auction for market clearance. Markets for access to data sources based on monthly subscription is another research direction to be investigated.

Finally, the adopted approach of IoTFeds will inspire the CoGNETs[9] project, as it shares some common principles with the emerging paradigm of Dataspaces [15] such as data sovereignty [11]. Hence, as a future work, the combination of the sophisticated marketplace mechanisms of IoTFeds (such as reputation system, pricing schemes, advanced search etc.) with Decentralized Identifiers (DIDs) and well-established and well-structured Dataspaces reference architectures such as GAIA-X [17] and IDSA [2] for the secure data exchange between entities could be a possible interesting research direction.

---

[9] https://www.cognets.eu/.

**Acknowledgment.** This work is partially funded by the IoTFeds project, co-financed by the European Regional Development Fund of the European Union and Greek national funds through the Operational Program Competitiveness, Entrepreneurship and Innovation, under the call RESEARCH - CREATE - INNOVATE (project code: T2EDK-02178) and by the European Union's Horizon Europe Research and Innovation Programme through the CoGNETs project under Grant Agreement No 101135930.

# References

1. https://www.i3-market.eu/. Accessed June 2024
2. IDSA reference architecture model 3.0. International Data Spaces Association (2021)
3. Athanasakis, E., et al.: Trustworthy decentralized management and governance of internet of things data federations. In: IEEE IoTaIS, pp. 247–253. IEEE (2023)
4. Azad, M.A., et al.: Decentralized self-enforcing trust management system for social internet of things. IEEE IoT-J **7**(4), 2690–2703 (2020)
5. Bauer, D.P.: ERC-20: Fungible Tokens. In: Getting Started with Ethereum: A Step-by-Step Guide to Becoming a Blockchain Developer, pp. 17–48. Springer, Berkeley (2022)
6. Diaz, F., et al.: Evaluating stochastic rankings with expected exposure. In: Proceedings of the 29th ACM CIKM, pp. 275–284 (2020)
7. Jacoby, M., et al.: D2.1 - Semantics for IoT and Cloud Resources, June 2017. https://doi.org/10.5281/zenodo.817469
8. Jacoby, M., et al.: D2.4 - Revised Semantics for IoT and Cloud Resources, July 2017. https://doi.org/10.5281/zenodo.827229
9. Kollmann, T., et al.: Toward a renaissance of cooperatives fostered by blockchain on electronic marketplaces: a theory-driven case study approach. Electron. Mark. **30**(2), 273–284 (2020)
10. Lawrenz, S., Sharma, P., Rausch, A.: Blockchain technology as an approach for data marketplaces. In: Proceedings of ICBCT, pp. 55–59 (2019)
11. Mertens, C., et al.: A framework for big data sovereignty: the European industrial data space (EIDS). In: Curry, E., Scerri, S., Tuikka, T. (eds.) Data Spaces, pp. 201–226. Springer, Cham (2022). https://doi.org/10.1007/978-3-030-98636-0_10
12. Monrat, A.A., Schelén, O., Andersson, K.: A survey of blockchain from the perspectives of applications, challenges, and opportunities. IEEE Access **7**, 117134–117151 (2019)
13. Nitti, M., Girau, R., Atzori, L.: Trustworthiness management in the social internet of things. IEEE TKDE **26**(5), 1253–1266 (2013)
14. Pathak, A., Al-Anbagi, I., Hamilton, H.J.: TABI: trust-based ABAC mechanism for edge-IoT using blockchain technology. IEEE Access (2023)
15. Scerri, S., Tuikka, T., de Vallejo, I.L., Curry, E.: Common European data spaces: challenges and opportunities. In: Data Spaces: Design, Deployment and Future Directions, pp. 337–357 (2022)
16. Sharma, P.S., Yadav, D., Garg, P.: A systematic review on page ranking algorithms. Int. J. Inf. Technol. **12**(2), 329–337 (2020). https://doi.org/10.1007/s41870-020-00439-3
17. Tardieu, H.: Role of Gaia-X in the European data space ecosystem. In: Otto, B., ten Hompel, M., Wrobel, S. (eds.) Designing Data Spaces, pp. 41–59. Springer, Cham (2022). https://doi.org/10.1007/978-3-030-93975-5_4

18. Tkachuk, R.V., et al.: A survey on blockchain-based telecommunication services marketplaces. IEEE TNSM **19**(1), 228–255 (2021)
19. Wang, S., et al.: Decentralized autonomous organizations: concept, model, and applications. IEEE Trans. Comput. Soc. Syst. **6**(5), 870–878 (2019)
20. Xie, J., et al.: A survey of blockchain technology applied to smart cities: research issues and challenges. IEEE Commun. Surv. Tut. **21**(3), 2794–2830 (2019)
21. Zarko, I.P., et al.: Towards an IoT framework for semantic and organizational interoperability. In: GIoTS, pp. 1–6 (2017)

# Edge Computing and Cross-Domain Systems

# Towards a Functional Continuum Operating System – ICOS MetaOS

Artur Jaworski[1](✉) [iD], Marcin Kotlinski[1] [iD], Izabela Zrazinska[2] [iD],
Xavier Masip-Bruin[3] [iD], Jordi Garcia[3] [iD], and Iman Esfandiyar[4] [iD]

[1] Poznan Supercomputing and Networking Center, Jana Pawla II 10st, 61139 Poznan, Poland
arturj@man.poznan.pl
[2] Worldsensing, Viriat 47 10th Floor, 08014 Barcelona, Spain
[3] UPC BarcelonaTECH CRAAX Lab, Rambla Exposicio 59, 08800 Vilanova i la Geltru, Spain
[4] Lukasiewicz Research Network–Poznan Institute of Technology, Ewarysta Estkowskiego 6 st, 61755 Poznan, Poland

**Abstract.** This article presents the intermediate results of the ICOS project, which aims to create a meta-operating system for cloud continuum, and describes the adaption of ICOS in two (out of four) project's pilot Use Cases. For each described scenario, the paper provides insight into the specific key architectural features, identifies the expected benefits, and provides some initial validation results. The purpose of this research is to highlight the advantages of using a metaOS in real scenarios as well as to illustrate the ease of integration into ICOS.

**Keywords:** ICOS · Continuum · MetaOS · Internet of Things

## 1 Introduction

### 1.1 ICOS as an Implementation of Cloud Continuum

The cloud continuum has been defined as an extension of the traditional cloud towards multiple entities, such as edge, fog, and the IoT, providing analysis, processing, storage, and data generation capabilities [1]. The continuum leverages the combined benefits of both computing paradigms -in short, unlimited resources and performance at the cloud, and low latency, reduced network usage, and increased data privacy and security at the edge.

An effective management of the continuum is a complex task and poses several research challenges that need to be addressed. For instance, managing a large number of heterogeneous, highly dynamic and mobile devices, considering the distributed nature of the different data assets as well as their privacy and security constraints, selecting the appropriate set of nodes where a multi-component application should be deployed for an efficient execution, or considering interoperability and interdependency between the various virtualization technologies of the running nodes, just to name a few.

Resources management and container orchestration at the cluster level is a relatively mature topic with several well-known products on the market, such as Kubernetes [2],

© The Author(s) 2025
M. Presser et al. (Eds.): GIECS 2024, CCIS 2328, pp. 119–133, 2025.
https://doi.org/10.1007/978-3-031-78572-6_8

Docker Swarm [3] or OpenShift [4]. However, extending the scope of management and orchestration to multi-cluster and heterogeneous edge environments becomes a more challenging topic of current research, which is attracting attention of a large number of researchers, academics and tech developers.

ICOS [5] is a research project that aims to design, develop, and validate a meta-operating system for the Cloud-Edge-IoT continuum by addressing four main challenges:

- device heterogeneity and volatility, continuum infrastructure virtualization and diverse network connectivity,
- optimized and scalable application execution and performance, including resource consumption, guaranteed trust, security and privacy,
- reduction of development and integration costs,
- effective mitigation of cloud provider lock-in effects.

Efforts towards these objectives will be consolidated in a data-driven system built upon the principles of openness, adaptability, data sharing and a future edge market scenario for services and data orchestration.

## 1.2 ICOS Architecture

ICOS has been conceived as a dynamic and elastic metaOS platform distributed across the continuum. The primary tasks of ICOS are twofold: a detailed management of heterogeneous resources in real-time, and an efficient deployment of applications on multiple nodes along the continuum. The ICOS design has been defined through two different roles: the ICOS Controller and the ICOS Agent. On the one hand, the ICOS Controller is responsible for managing the continuum (tracking the current system topology and availability) and the run-time (deploying and monitoring application execution on demand). On the other hand, the ICOS Agent is responsible for executing the offloaded users' applications, taking care of code execution, data access, telemetry collection and, eventually, runtime communication with other Agents. There is an ICOS Agent running on each node of the continuum (whether a complex cluster or a constrained device at the edge) and it is the only ICOS software running on the remote infrastructure.

ICOS has been designed as a distributed, multi-controller system, in which all relevant decisions are made at the Controller level. ICOS Agents receive instructions from ICOS Controllers and translate them into infrastructure-specific commands. With this organization, ICOS is shaped as a technologically independent platform, where abstract decisions are made without technological restrictions (at the Controller level) and are implemented taking advantage of the capabilities of remote infrastructure technologies (at the Agent level). A typical ICOS scenario is illustrated in Fig. 1.

ICOS Controllers are distributed along the continuum to leverage locality, providing fast response time and supporting scalability. ICOS Agents are deployed at every node of the continuum. They can range from powerful clusters to constrained computing devices and provide processing and data accessing capabilities. Each ICOS Controller is in charge for managing a number of Agents based on locality principles. Upon request for application execution, the Controller will attempt to find locally an appropriate set of nodes within its scope to execute the application according to the application requirements; in case the request cannot be satisfied with the available resources, the Controller

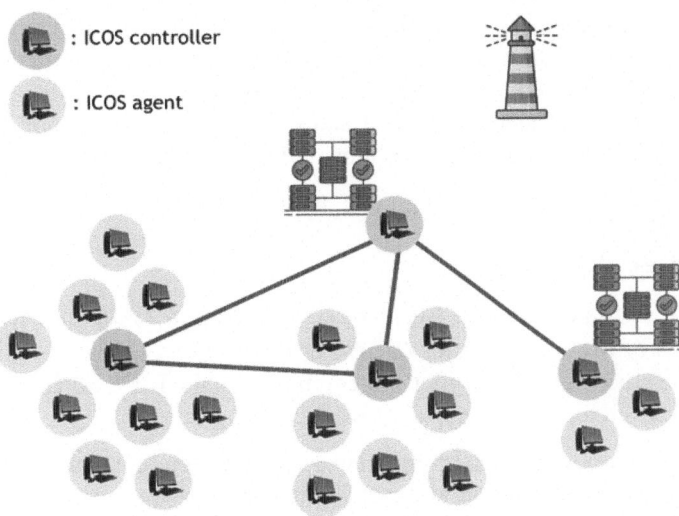

**Fig. 1.** Typical ICOS scenario.

will coordinate horizontally with other Controllers to find a global solution for the application request.

The ICOS Controller has been designed as a three-layer architecture. As shown in Fig. 2, the Meta-Kernel layer implements all tasks related to the continuum management, as well as the runtime decision making and management. This layer is also responsible for collecting telemetry data from the infrastructure through the Agents. The Intelligence layer is fed by the telemetry data, and it is the responsible for providing intelligence to the Meta-Kernel layer (both for the continuum and runtime management decisions) as well as providing predictive monitoring to forecast different runtime events (resource

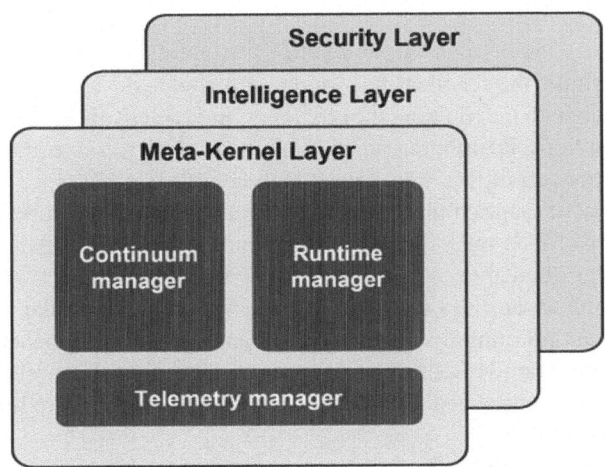

**Fig. 2.** Architecture of the ICOS Controller.

utilization, network load, or security risks, among others). This layer is also responsible for launching training and retraining processes through federated learning to keep the intelligence model updated. And finally, the Security layer is responsible for providing identity and access management, trusted and encrypted communication channels, as well as detecting security related risks (vulnerabilities, threats, or anomalies) and events (audit).

### 1.3 Objectives

In this paper, we present the intermediate results of the ICOS project by describing the integration of two real scenarios (as part of the ICOS project's Use Cases [6]) and discussing the challenges in the pilot's integration with the ICOS Alpha release. ICOS has been conceived to face the following main challenges:

- enabling technology agnostic operation in a heterogeneous continuum infrastructure,
- facilitating an on-demand ad-hoc and AI-assisted development of the continuum infrastructure,
- ensuring reliability and security of the continuum,
- creating an open platform facilitating resources, models, data and services sharing, promoting EU innovation and new business models in the continuum arena [7].

The paper is organized as follows. After describing the main architecture and features of ICOS, in Sect. 2 and Sect. 3 we present the two project's pilot use cases, in the areas of agriculture and railway system, respectively. For each pilot, we describe the specific architecture, the expected benefits of using ICOS, the integration with the Alpha release, and the next steps. And finally, Sect. 4 discusses the advantages of using ICOS and concludes the research.

## 2 Project's Pilot Use Case 1 Perspective

### 2.1 Overview

In modern agriculture, the seamless flow of data from edge devices to cloud-based platforms is fundamental to maximizing the efficiency and effectiveness of robotic and technological interventions. Distributed processing, data transfer, connectivity, and security are integral components of this data-driven approach [8] (Fig. 3).

At the edge of the agricultural system, sensors, drones, and robotic devices collect vast amounts of data on soil health, crop conditions, weather patterns, and more. These edge devices often operate in remote areas with limited connectivity, necessitating onboard processing capabilities to analyze data in real-time and make immediate decisions. Edge computing minimizes latency and bandwidth requirements by processing data locally, enabling rapid response to changing conditions without relying on constant communication with centralized servers. However local processes are limited by device computational power [9].

The main task of the project at the current stage is to detect weeds and crops and localize them with a centimeter accuracy in global coordinates. This data is then used

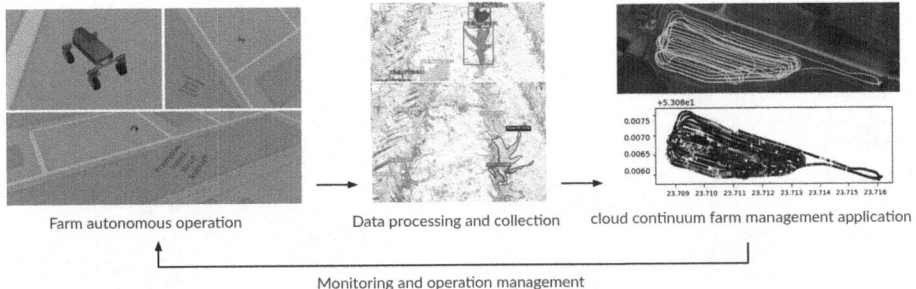

Farm autonomous operation          Data processing and collection     cloud continuum farm management application

Monitoring and operation management

**Fig. 3.** Use Case's main functionality.

to generate a map of the field containing crop and weed intensity. Once data is collected and preprocessed at the edge, it needs to be transferred to cloud-based platforms for further analysis, storage, and integration with other data sources. High-speed, reliable connectivity solutions, such as cellular networks or satellite communication, facilitate the transmission of data from remote agricultural sites to cloud servers.

In the cloud, agricultural data undergoes extensive analysis using advanced analytics, machine learning, and artificial intelligence algorithms. These tools extract valuable insights from raw data, identifying parameters that inform decision-making processes. For example, predictive analytics models can forecast crop yields, diseases, or optimal planting times based on historical data and current environmental conditions.

Managing the comprehensive system of agricultural data flow from edge to cloud, including data transfer, distributed processing, and storage, poses a complex challenge that often falls beyond the expertise of agricultural software developers. To address this challenge, leveraging an infrastructure capable of supporting such an environment becomes essential. The main challenge to be addressed by integrating with ICOS is providing continuous robot operation in rural areas with low network coverage and high latency.

## 2.2 Architecture

Use Case's architecture spreads across cloud, edge and IoT domains. Depending on the mission, the robot can take advantage of cloud computing, for instance, detecting and localizing weeds and crops based on AI models. The data generated during the mission, along with the status and record of the operation in the field, is sent to cloud infrastructure. The cloud processes analyze the farm condition and robot health and generate a farm yield map. This generated data is then sent to the robot as the next task and mission to be executed on the farm. For instance, the robot might execute a spraying mission at the part of the farm where weak crop conditions were previously detected.

**Hardware Specification.** The robotic platform, shown in Fig. 4, has mechanical tools for autonomous seeding and spraying, equipped with a hydraulic power drive system and a diesel engine. The robot takes advantage of four active steerable wheels equipped with an independent suspension system for a smooth maneuver on the farm (Fig. 5).

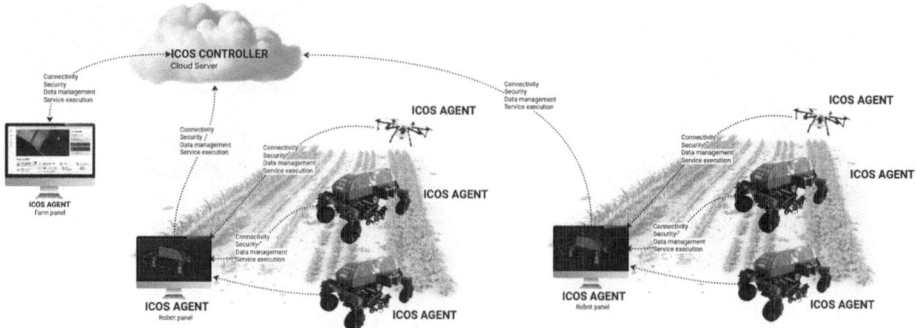

**Fig. 4.** Use case architecture overview.

**Fig. 5.** AgroRob, the agricultural robot used in this work.

A 24 V designated battery pack connected to the alternator of the diesel engine, powers the high-performance onboard computer. The robot's CAN bus is connected to this computer via USB adapter. To establish connectivity, an industrial 5G router is connected via LAN port.

The robot is localized by GNSS receivers, as well as odometry data received from wheels' encoders [10]. Three RGB cameras and a depth camera are mounted on the robot as the vision sensors. Robot Operating System version 2 (ROS2) [11] has been used as an interface between robot and the computer, while You Only Look Once, version 8 (YOLO8) [12] is responsible for image recognition.

### 2.3 Expected Benefits

End User added value for the Use Case implementation:

- enhanced operational efficiency: autonomous robots perform precise tasks such as seeding, weeding, and spraying, reducing the need for manual labor and increasing task accuracy,
- cost reduction: predictive maintenance and optimized resource application lead to significant savings on fertilizers, herbicides, and repair costs,

**Fig. 6.** Hardware setup, including communication hardware and sensors

- sustainability: efficient resource use and reduced chemical application promote environmentally friendly farming practices,
- improved decision-making: real-time data and analytics support informed decisions, optimizing crop management and increasing yield.

Applications Executed with ICOS Support and Requirements (Challenges) for ICOS:

- Autonomous Field Monitoring: this application utilizes the robot to autonomously monitor crop conditions, soil health, and environmental factors. The platform, equipped with advanced sensors and cameras, collects real-time data to enable precise interventions, enhancing crop health and yield. The application requires ensuring reliable integration of various sensors, managing large volumes of data, and providing actionable insights in real-time. There is also a need for robust connectivity solutions such as LoRa [13], WIFI, and xG modems to facilitate data transmission.
- Weed and Disease Detection: leveraging machine learning and computer vision, this application identifies weeds and diseases early, enabling targeted treatments. The system processes data locally on the edge to ensure timely responses even with intermittent connectivity. The weed map generated during the robot's first pass helps in precision treatment during subsequent passes. The application requires developing accurate detection algorithms, managing the computational load for processing image data on the edge, and maintaining high performance under varying field conditions.
- Predictive Maintenance and Resource Optimization: this application uses predictive analytics to anticipate machinery maintenance needs and optimize resource usage. By analyzing sensor data and operational logs, it schedules maintenance proactively and adjusts resource application to prevent overuse. Data from cameras, logs, and other devices are stored in the cloud, with predictive analysis considering vibrations and control signals. The application requires creating robust predictive models, balancing computational loads between edge and cloud resources, ensuring data security, and

developing user interfaces for maintenance management and parameter control. There is also a need for securing the connection due to the high value of robotic devices.

## 2.4 Integration with the ICOS Alpha Release

Once the requisite hardware has been installed, the iterative implementation of ICOS could commence. As all the operating systems used in Use Case support Docker, the next step was to select an orchestrator. Nuvla [14] has been selected for this purpose. Nuvla serves as a comprehensive platform that facilitates the seamless deployment, monitoring and scaling of "dockerized" applications, thereby increasing the efficiency and reliability of containerized environments.

With all these preparations completed, the actual onboarding could begin. A virtual machine, located on PSNC premises and hosting the Zenoh router, was selected for initial onboarding. The Edge component was registered in Nuvla and subsequently deployed on the Docker engine running on the virtual machine, utilizing the provided docker-compose YAML files. Following the completion of the deployment, the Edge registered in Nuvla underwent a transition to the operational status within a few seconds. The procedure was executed without incident, in accordance with the instructions provided in the documentation.

The final stage of the process involved the deployment of the ICOS Telemetry Agent on the virtual machine that had been previously onboarded and the subsequent connection of this agent to the Telemetry Controller, which was running on the staging testbed for Use Cases integration. This final step may be divided into two sub-steps: firstly, establishing a connection to the testbed via a VPN client, and secondly, initiating the Telemetry Agent. VPN client was deployed on the virtual machine, thereby establishing a connection to the testbed. Following the implementation of a series of corrective measures, the deployment was successfully concluded, resulting in the commencement of data reporting by the Telemetry Controller, as evidenced by the Fig. 6. Both the VPN client and the Telemetry client have been deployed by Nuvla (Fig. 7).

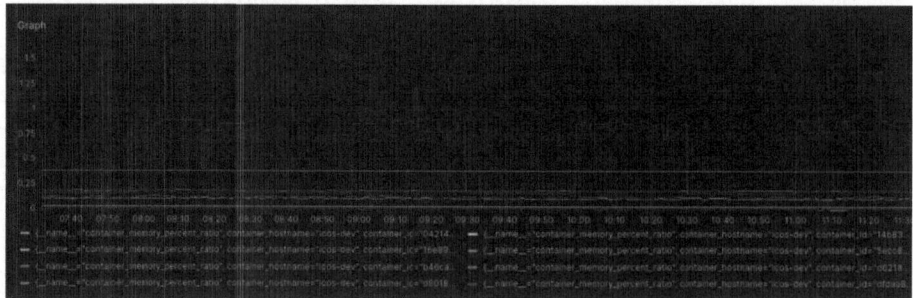

**Fig. 7.** Telemetry data from UC1 virtual machine being reported by the Telemetry Controller.

## 2.5 Next Steps

The subsequent stages of integration into the ICOS environment will involve preparation for the Beta release. The main point of interest for the Use Case will be validation of own applications deployment using ICOS on both the cloud and the robot. For this to happen, Docker Compose YAML files will have to be translated into ICOS Application manifest format.

After successful deployment, the work on leveraging ICOS's Intelligence Layer will commence. It is expected to be a base for creating the aforementioned Predictive Maintenance module, as well as help in training more accurate image recognition models used on the robot.

Apart from work directly connected to using ICOS, other modules will also be revamped. The team will focus on creating user dashboards for web and mobile devices. These applications will allow users to plan missions, see their results, as well as monitor the robot's parameters in real-time.

## 3 Project's Pilot Use Case 2 Perspective

### 3.1 Overview

Ferrocarrils de la Generalitat de Catalunya (FGC) infrastructure includes metro and commuter lines in and around the city of Barcelona, tourist mountain railways, and rural railway lines which serve more than 90 million passengers per year. Whilst most lines are conventional adhesion railways, the FGC also operates two rack railways and four funicular railways.

On all these railway lines, the massive deployment of sensors along different parts of the infrastructure is essential for the optimization and improvement of service and safety. The increasing number of sensors and their specific, and typically siloed solutions, present an increasing complexity related to the management and operations of such solutions.

Today, the railway monitoring process to improve the maintenance cycle is basic, and for most railway operators it is done preventively (once every fixed period) through a special train with sensors which runs through the whole rail system [15]. This special train can measure several key parameters of the railway system, such as the height difference and width between the rails, and thus identify where, potentially, corrections in the track geometry is needed. However, this measurement is only taken once or twice a year; in the remaining months, nobody knows what happens (only physical inspections are available: very costly and uncommon), and there is no established procedure to evaluate the cost-effectiveness of the actions taken to address the identified rail tracks' issues. Indeed, digital technology, such as IoT, aims to minimize the monitoring and maintenance costs by gaining knowledge of the status of key aspects of the railway infrastructure in real-time: rail tracks geometry, slope, surrounding areas settlements and falling elements, overhead lines maintenance, etc.

The main challenge to be addressed by the Use Case is related to the continuous monitoring of critical infrastructure on rail tracks to ensure safety and improve maintenance activities.

The initial area to deploy and validate the Use Case for the Railway Structural Alert Monitoring system (RSAM) is the line in Lleida-La Pobla due to its difficult access to several of the areas of the line and its orography generating possible geological incidents (Fig. 8).

**Fig. 8.** Safety problem on the FGC rail track.

## 3.2 Architecture

In November 2023 (M15), as part of Use Case 2, a multitude of IoT devices were strategically deployed along the FGC rail tracks. The deployment site spans the Lleida-La Pobla line, covering a 4 km stretch within the challenging terrain of Gerb. This particular rail line, facilitating 16 train circulations daily, stands as the sole transportation artery for the region. Characterized by its precarious geodesic conditions, including water flow beneath the tracks and non-compacted layers, as well as the presence of small caves and historical instances of ground collapse, the area underscores the critical need for a real-time monitoring solution. This solution, intended to significantly impact end-users, notably the railway operator FGC, is being implemented by Worldsensing (WSE) within this Use Case and is set for validation by ICOS (Fig. 9).

**Fig. 9.** IoT devices installed on the FGB railway track as a part of ICOS Railway Use Case

For the Railway Structural Alert Monitoring system, ICOS will be managing the Edge and Cloud processing environments. Edge will be supported by the IoT Gateway with limited resources for computing and 4G connectivity through commercial mobile services to the Cloud computing environment. The cloud computing environment used by Worldsensing is provided by Google Cloud Platform. Both the Edge device and the Cloud environment should have the ICOS agent deployed to be able to onboard such elements to the continuum.

The onboarding of both compute services will allow the orchestration of services through ICOS Meta OS according to specific requirements for the Monitoring, Safety and Maintenance applications available in the CMT Cloud solution.

Out of the scope of the ICOS-managed environment, data from the IoT sensors and nodes will be aggregated at the IoT gateway through LoRaWAN radio communication. The data collected from the IoT sensors (tiltmeters) is related to the geometry parameters of the rail track, while IoT nodes collect data from geotechnical sensors (extensometers and piezometers) to geological parameters (Fig. 10).

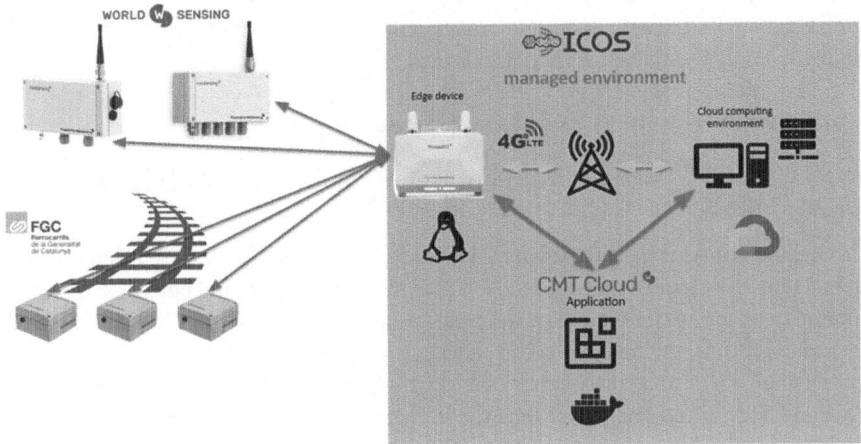

**Fig. 10.** Use Case 2 architecture.

### 3.3  Expected Benefits

End User added value for the Use Case implementation:

- timesaving through continuous monitoring: limit intensive personnel inspections that are done every day before train circulations,
- cost saving: implement corrective actions in advance to avoid reparation costs,
- improve safety: establish velocity limits to avoid risky situations when with quality of the operation decreases.

Applications executed with ICOS support and requirements (challenges) for ICOS:

- Real-time Monitoring: the application will support safe operations by deploying a digital and wireless monitoring system that will collect and deliver real-time information regarding the quality parameters to monitor critical infrastructure status to support the decision makers and to timely detect possible anomalies or physical threats regarding the railway track.

- Critical event detection for Safety: the alarm detection module is connected to the deployed devices, and it allows the detection and acquisition of possible alarms. The detection of alarms and response actions can be required to be processed at the edge to ensure safe operations even if connectivity with upper layers is not fully available. Such response to events detected might also include the request for additional information to the physical devices to collect additional contextual data about the possible incidents and thus to better design or select a response plan.
- Prediction for maintenance planning: to optimize the decision-making process and exploit all the available resources, the maintenance application will also be onboarded within the ICOS architecture. The objective of the application is to identify the trend and predict the moment when the condition where quality parameters would not be met, and therefore plan maintenance activities to mitigate such risk. The proposed application will request available resources at the edge and cloud level based on connectivity and data transfer requirements and will run appropriately.

### 3.4  Integration with the ICOS Alpha Release

To prepare the ICOS for its initial deployment, the WSE team tested different Gateway models, one of which is currently under development.

**Deploying Nuvla Agent on Gateway.**  The objective of the test was to be able to deploy a Nuvla agent in a Gateway, using Docker. The WSE team tried to keep track of the CPU and memory usage in the Gateway, to install more services (such as Telemetry or VPN Agent) in the future, the team has chosen to use htop, to monitor the state of the system. The use of a CPU with components that are needed for basic Gateway functionalities uses around 50% of its capacity. Once the team started the docker services, there was a

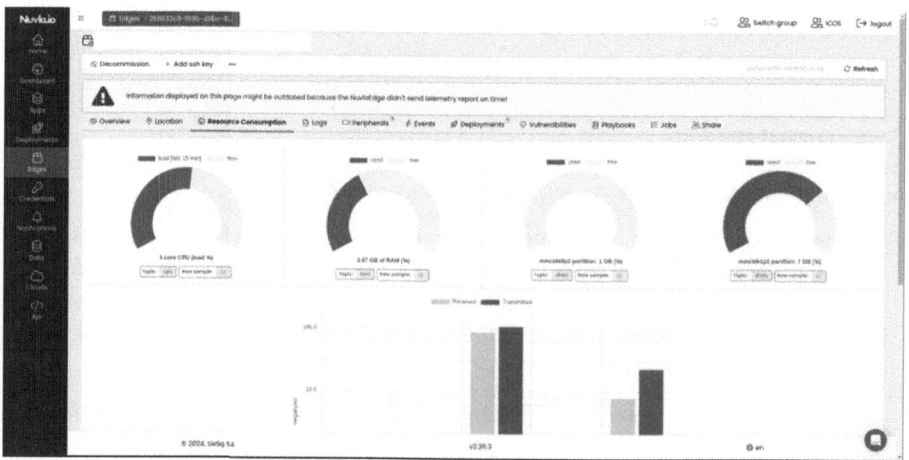

**Fig. 11.** Nuvla dashboard for UC2 resource consumption on the Worldsensing Gateway.

noticeable increase in CPU but the memory usage did increase, at least with a memory reservation. When the docker-compose images started to work the CPU usage increased clearly (>40%) (Fig. 11).

Based on this test, it was decided to deploy the Nuvla agent using Docker, instead of Kubernetes, as Kubernetes could not be configured satisfactorily and the CPU usage increased more than with Docker.

**Deploying Telemetry and VPN Agents via Nuvla.** The final step was to deploy Telemetry and VPN Agents remotely. It allowed to confirm that the device will be able to maintain its functionality. In further steps, some improvements will be made to maintain the GW and the system stable.

### 3.5   Next Steps

The next steps for the Use Case 2 ICOS setup involve collecting and adapting docker-composed architecture (Nuvla Agent) for edge application components installed on the Gateway and translating Kubernetes (k8s) manifests for cloud application components (CMT). These manifests will first be adapted to the ICOS manifesto template. Following successful testing, the deployment will be conducted through ICOS, with careful monitoring to ensure all components are operational. In later stages, UC2 might consider testing ICOS implementation with the OCM agent which could be installed and tested to verify its interactions with the cloud components.

Once Use Case 2 is fully onboarded into the ICOS system it is expected to focus on functionalities that will bring the impact to end user operations such as:

- real-time monitoring: data integrity and synchronization if there are connectivity problems between Edge-Cloud,
- critical event detection for safety: operate regardless of connectivity (taking local decisions),
- prediction for maintenance planning: identify the trend and predict the moment when quality parameters would not be met.

By the end of the project, we aim to confirm the business value that the use of the ICOS system brings to transportation. This includes enhanced safety and efficiency due to real-time data collection and edge processing, which allow for immediate response to potential issues. The system facilitates cost savings through predictive maintenance and optimized resource allocation, while also improving passenger satisfaction by reducing delays and enhancing service reliability. Compliance with regulatory standards is maintained through precise, up-to-date data, and advanced asset management is enabled by comprehensive cloud analytics, optimizing railway operations and infrastructure management.

## 4   Conclusions

As presented, ICOS strives to create an easy to integrate solution, which will enable usage of the edge-to-cloud continuum by its adopters. All the modules, which compose the architecture, as well as their features, are iteratively consulted with the early adopters

- four project's pilot Use Cases (two of which were presented in this document), as well as organizations selected in two Open Calls (the second one is open for applications at the time of writing this document). Choosing ICOS architecture as a base for developed solutions will help minimize their time to market, improve reliability and open new possibilities for maximizing benefits of using cloud, edge and IoT domains in a form of a continuum. Our future efforts will be focused on validating the new features and providing feedback during the ICOS development.

**Acknowledgments.** The ICOS project has received funding from the European Union's HORIZON research and innovation programme under grant agreement No 101070177.

**Disclosure of Interests.** The authors have no competing interests to declare that are relevant to the content of this article.

# References

1. Moreschini, S., Pecorelli, F., Li, X., Naz, S., Hästbacka, D., Taibi, D.: Cloud continuum: the definition. IEEE Access **10**, 131876–131886 (2022)
2. Kubernetes Homepage. https://kubernetes.io. Accessed 20 June 2024
3. Docker Swarm Homepage. https://docs.docker.com/engine/swarm. Accessed 20 June 2024
4. OpenShift Homepage. https://www.redhat.com/es/technologies/cloud-computing/openshift. Accessed 20 June 2024
5. ICOS Homepage. https://www.icos-project.eu/. Accessed 20 June 2024
6. Zrazinska, I.: ICOS. Deliverable 6.4: Use Cases settings and demonstration strategy (2023)
7. D'Andria, F.: ICOS. Deliverable 2.1: ICOS ecosystem: technologies, requirements and state of the art (2023)
8. Debauche, O., Mahmoudi, S., Manneback, P., Lebeau, F.: Cloud and distributed architectures for data management in agriculture 4.0: review and future trends. J. King Saud Univ. - Comput. Inf. Sci. **34**(10), 9622–9643 (2022)
9. Tzounis, A., Katsoulas, N., Bartzanas, T., Kittas, C.: Internet of Things in agriculture, recent advances and future challenges. Biosys. Eng. **164**, 31–48 (2017)
10. Esfandiyar, I., Ćwian, K., Nowicki, M.R., Skrzypczyński, P.: GNSS-based driver assistance for charging electric city buses: implementation and lessons learned from field testing. Remote Sens. **15**, 2938 (2023)
11. ROS Homepage. https://www.ros.org/. Accessed 20 June 2024
12. YOLO8 Homepage. https://yolov8.com. Accessed 20 June 2024
13. Mikhaylov, K., Petäjäjärvi, J., Hänninen, T.: Analysis of capacity and scalability of the LoRa low power wide area network technology. In: European Wireless, pp. 119–124 (2016)
14. Nuvla Homepage. https://nuvla.io. Accessed 20 June 2024
15. Budai, G., Huisman, D., Dekker, R.: Scheduling preventive railway maintenance activities. J. Oper. Res. Soc. **57**(9), 1035–1044 (2006)

# Edge-Cloud Solution Based on FIWARE and Context Augmentation to Monitor Usages of Carpooling Car Parks

Gilles Orazi[1]([✉]) [iD], Marianne Marot[1] [iD], Iheb Khelifi[1] [iD], Franck Le Gall[1] [iD],
Amara Richard[2], and Yvan Martzluff[2]

[1] EGM, AREP Center, Traverse des Brucs, 06560 Valbonne, France
{gilles.orazi,marianne.marot,iheb.khelifi,franck.le-gall}@egm.io
[2] Conseil Départemental des Pyrénées Orientales, Perpignan, France
{amara.richard,yvan.martzluff}@cd66.fr

**Abstract.** Monitoring the usage and carbon impact of infrastructures such as carpooling car parks may be a challenge for local authorities. In this study, we address this issue by proposing an IoT solution that integrates both Edge processing and Cloud computing. Real-time continuous video AI processing automatically detects and stores information of car entries and exits. Then Cloud computing stores the anonymized data into a FIWARE information broker using ETSI's NGSI-LD specification and employs query language to process and display useful statistics about carpark usage. To understand the carpooling user behaviors for each car park, as well as the various Key Performance Indicators and threshold values that direct our monitoring and alerting systems, we analyzed 1 year of data acquisition and performed in-situ surveys. Hereby, we present our results that reflect two different usages of carpooling (professional and personal) in terms of stay durations, car fidelity, car energy distributions, and carbon impact. This solution was deployed and tested on 3 car park lots, 2 of which have been permanently running for about two years.

**Keywords:** Monitoring · Mobility · Outlier detection · Data Quality · Decarbonation · GHG emission reduction · FIWARE · NGSI-LD

## 1 Introduction

Handling mobility in the city in a sustainable way is a challenge that local authorities are tackling by encouraging new practices associated with (sometime new) specific infrastructures. The evaluation of their effectiveness compared to the policy objectives is a necessity to analyze behavior changes as well as global effects and spillovers. This is the case for car parks dedicated to carpooling. This practice is often used for commuting and is encouraged by local authorities because it leads to less vehicles on the roads, and less greenhouse gases emissions.

The design of carpooling car park is aligned with the aims of the European "Green Deal" which targets Carbon neutrality by 2050 and the achievement of three goals:

© The Author(s) 2025
M. Presser et al. (Eds.): GIECS 2024, CCIS 2328, pp. 134–150, 2025.
https://doi.org/10.1007/978-3-031-78572-6_9

- Ecological transition: To build carpooling car parks along isolated roads.
- Energy transition: To reduce Carbon Dioxide ($CO_2$) emissions.
- Digital transition: To produce and share data[1] to monitor and promote usage of carpooling and surroundings.

The *Conseil Départemental des Pyrénées Orientales* is an administrative department in Southwestern France which is constructing such carparks and has the necessity to comply to the Digital Transition by monitoring their usages, attendances and to estimate their positive $CO_2$ impacts in accordance with the Energy and Ecological Transitions. Since no solution yet existed on the market for such monitoring, we decided to implement a demonstrator at the *Perpignan Nord* car park and later to deploy this solution permanently to two other carpooling car parks: *Perpignan Sud* and *Mas Sabole* car parks.

In this paper, we describe our monitoring solution deployment experimentally at these three pilot carpooling carparks, located in the Southwestern France, near the conurbation of Perpignan. Our aim is to better understand their carpooling usages, attendances, and patterns of behaviors, as well as to estimate their positive $CO_2$ impact throughout time. Real-time monitoring is performed with displayed analytics on dashboards. Throughout these use cases, we demonstrate the feasibility of integrating Edge video monitoring, completed with offline data augmentation (data querying and anonymization), and running a data quality processing pipeline on a Cloud instance where our database is built. This solution enables us to overcome data privacy and security concerns and to be compliant with European RGPD regulations.

We identified in the literature only one research that tackled this same problem. It is reported in [1] and [2]. The authors report how they collected car park data usage using an automated license plate reader (ALPR) and how they used it to infer the various categories of users and managed to forecast the parking demand. Our solution is quite close to the methods of data collection and analysis but is more oriented toward an interoperable solution that can be implemented as a Smart City application.

## 2 Pilot Sites

We performed our experiments in three carpooling car parks, at different times. All are located near important highway interchanges as shown in Fig. 1.

**Perpignan Nord.** The first pilot use-case for our solution. It was deployed from November 2020 to March 2022, encompassing the first period of COVID-19 lockdown and the subsequent population release. Results have been presented in [3]. Here, video monitoring is performed with a single camera, whereby the entrance and exit are very near to each other. This pilot site is only presented here briefly to support the validation of our solution, however, is not used in this study.

**Perpignan Sud.** The second pilot use-case for our solution, building upon our experience from Perpignan Nord car park. It has been deployed since June 2022 and is still running today. Here, video monitoring is performed with two cameras, each oriented towards the entrance and exit lanes, respectively.

---

[1] Open data program: http://transport.data.gouv.fr.

**Mas Sabole.** The third pilot use case for our solution, is deployed since December 2022 and still running today. As for Perpignan Nord, here, video monitoring is also performed with a single camera, whereby the entrance and exit are very near to each other.

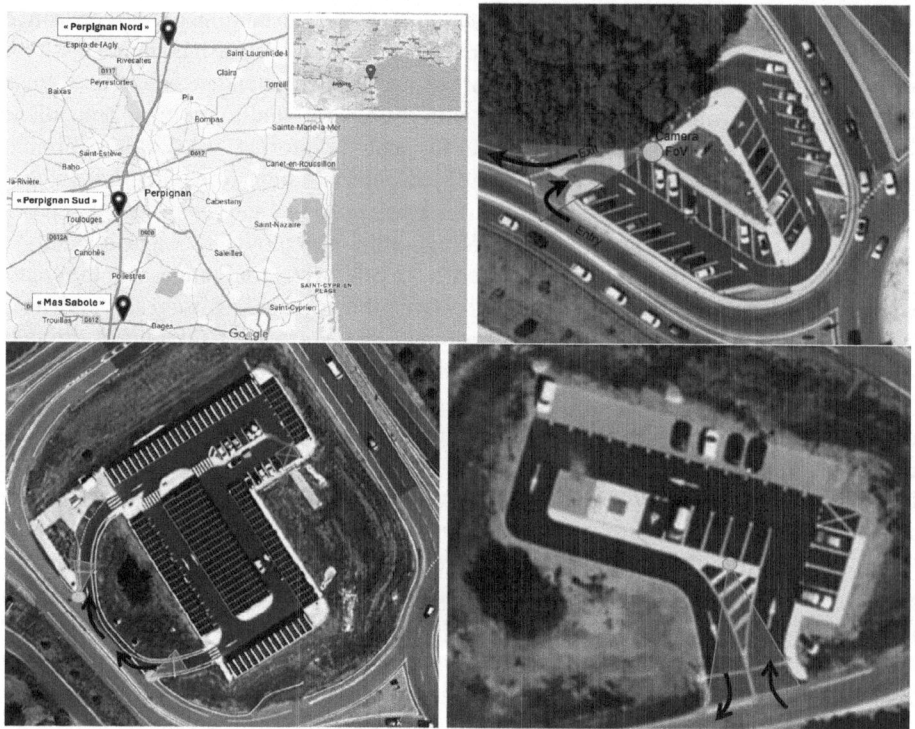

**Fig. 1.** Map view (upper left) and aerial views of the carpooling carparks: Perpignan Nord (upper left), Perpignan Sud (lower left) and Mas Sabole (lower right). Purple cones represent the cameras' lines of sight. (Color figure online)

## 3   Architecture and Implementation of the Solution

To meet our statistical goal, we need to identify each vehicle and to record precisely the time of their entrance and exit of the car park. Moreover, we want to ensure the users privacy to comply with the GDPR and that the solution is interoperable with other smart city applications. This section describes how this was implemented.

The solution we designed comprises various processing modules which are depicted on Fig. 2. We classified them as: edge processing, data reception, context augmentation, and statistics plus dashboarding. The edge processing hardware and software parts are located in the car park, from which the data are sent using a 4G connection to the data reception module hosted in the cloud. All the processing done on the edge side is intended to lower the bandwidth on the 4G network and to ensure the privacy of the users. In one word, that mainly means that no image is neither transmitted through the network nor saved anywhere. The following sections describe these modules in more detail.

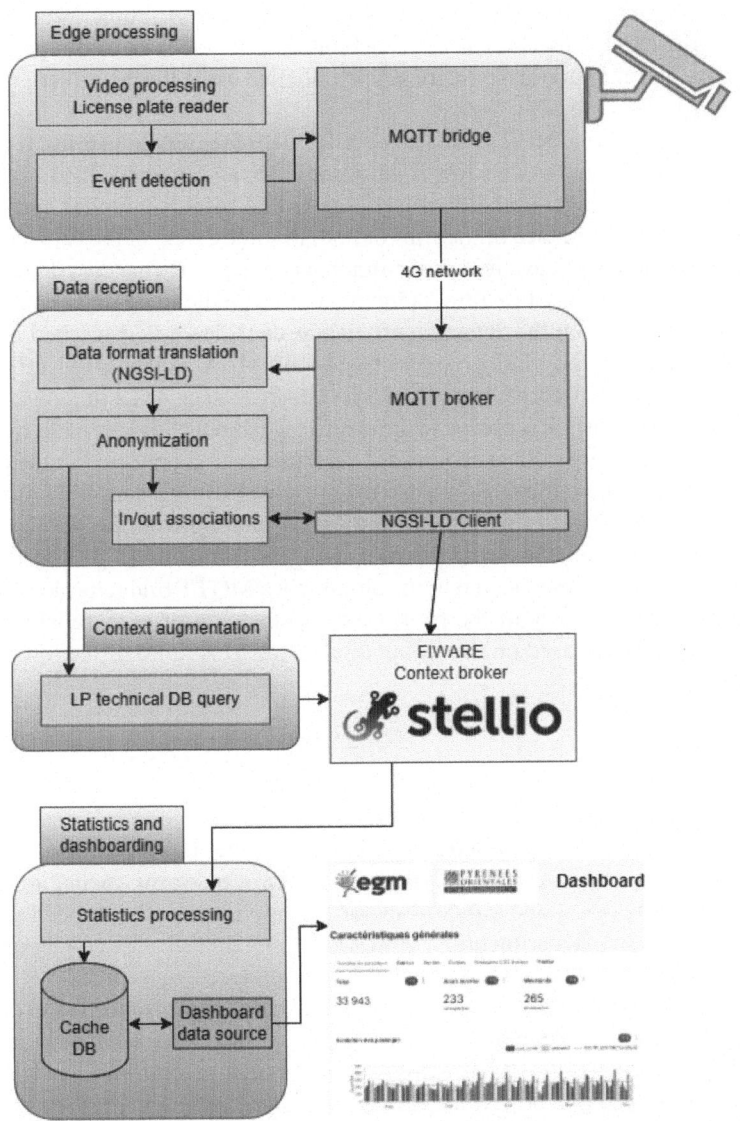

**Fig. 2.** Global architecture of our solution for monitoring carpooling carparks.

## 3.1  Edge Deployment

The deployed hardware in each car park is a camera and an edge computer. The setup at the Perpignan Nord car park was the first version of the system. It has been described in

[3] and has slightly changed since then. The setup deployed in the Perpignan Sud and Mas Sabole car parks is almost the same. Some cameras with license plate detection capabilities are installed in front of the access of the car park. In the Perpignan Sud there are two of them, one for the entrance and another for the exit. In the Mas Sabole, the entrance and the exit are located close to each other so that they can be monitored by a single camera. The entrance and exit are simply labelled using the position of the license plate in the picture.

The camera is the model IDS-2CD7A46G0/P-IZHS(8-32MM) from Hikvision. It was mainly chosen because of its low light capabilities and its embedded licence plate recognition system. It is mounted on top of a pole at a sufficient height above ground such that it can see the entrance or the exit (or both in some cases) of the cars. One should notice that these car parks have no barrier and that consequently the cars do not stop for a short time when entering of exiting, inducing a more challenging detection task. This has some consequences on the detection efficiency, considered as described in part 5.

The Edge computer is a Modberry-500-CM40408-MAX which is basically a rugged Raspberry Pi with a Quad-core Cortex-A72 1.5 GHz Processor, 4 GB of RAM, and 8 GB of eMMC Industrial Flash. It is placed in the electrical cabinet of the car park, from which the power is used to power the installation. The cameras are connected to the computer by means of a long ethernet cable that also brings power using a simple POE injector.

The edge computer collects the detection events coming from the cameras, it then formats a JSON message containing vehicle passage information (Fig. 2) which is sent to the data reception module, hosted in the cloud, via a MQTT bridge/broker setup with QoS 1 to ensure the delivery of the packet even in the case of an unreliable network. The MQTT software is based on the installation of the Mosquitto broker on both sides of the bridge.

## 3.2  Cloud Processing

### NGSI-LD/FIWARE Ecosystem Presentation

The implementation of the solution discussed in this article should be considered within the framework of a smart territory platform. In such a platform, managing data is a significant challenge due to the influx from various applications developed by different individuals across city departments or cities. The anticipated advantages of a smart territory platform lie in the ability to integrate these diverse data sources.

This challenge inspired ETSI, an EU standardization organization, to establish the Industry Specification Group for cross-cutting Context Information Management (ISG-CIM). This group focuses on facilitating the *"exchange of information, with proper formal definitions, between vertical applications, so that these applications retain their original meaning"*. They publish a technical specification known as NGSI-LD, enabling multiple organizations to develop interoperable software for managing cross-cutting context information. This allows applications to discover, access, update, and manage context information from various sources and publish it through interoperable data platforms.

While having such a standard is beneficial, it is not sufficient alone because:

- The standard must be implemented and supported by various software modules necessary for deploying a real-life smart city software platform (e.g., data connectors, processing engines, data publication platforms, etc.).
- A set of standard data models is essential to simplify interoperability.
- An open-source software approach is ideal as it helps avoid vendor lock-in, achieve economies of scale in development, and retain data ownership.

This is the mission of the FIWARE foundation, which provides and promotes a suite of open-source software modules for smart solutions based on open standards. It specifically utilizes NGSI-LD for context management and offers a comprehensive and continually evolving set of standardized data models known as smart data models. FIWARE promotes various implementation of the NGSI-LD standard, and we have chosen to use the one called Stellio (that is also the one our company have implemented).

The NGSI-LD data model is entity-based [4]. An entity is understood to be an informational representation that is supposed to exist physically or conceptually. Links between entities is ensured by NGSI-LD relationships. Entities and Relationships are described via NGSI-LD properties.

We used this standard to maintain a digital twin of our car park, with the data model described on Fig. 3. It does not explicitly store the events sent from the edge side but is rather aimed at providing a list of Stay entities. Each of these entities represents a vehicle that had entered, stayed, and exited the car park. Each Stay is linked to a Vehicle and a Site (the car park). So, if a given vehicle was stationed 3 times on the same park, this will be modelled by 3 Stay entities in the Context Broker.

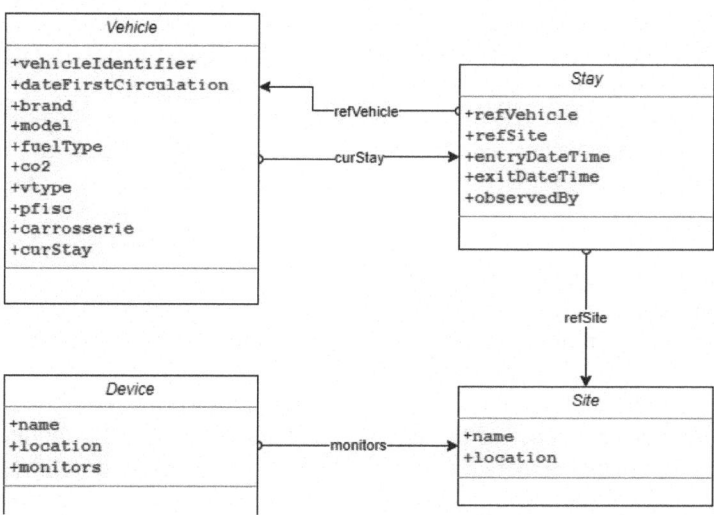

**Fig. 3.** NGSI-LD data model of the carpooling monitoring solution.

**Cloud Workflow**

Upon the arrival of an event, the data reception module updates the digital twin of the car park. The Stay entity is built in two steps upon the reception of the entry and exit events using the following workflow.

- When an entry event is received, a new Stay entity is created and linked to the Vehicle entity that corresponds to the detected license plate. If needed, a new Vehicle entity is created at this stage if it does not already exist for this license plate. At this stage, the Stay entity is incomplete and does not have any exit datetime. The Vehicle entity has a property that points to the current incomplete Stay, modelling the fact that this vehicle is known to be currently stationed in the car park.
- When an exit event is received, the incomplete Stay is retrieved by the means of the vehicle and updated by setting the exit datetime. At this stage, the Stay entity is complete and unlinked from the Vehicle, modelling the fact that the vehicle is no longer stationed in the car park.

When the Vehicle entity is created, a context augmentation is performed by querying the official SIV database which contains the technical information of all the registered vehicles in France.

The dashboards, allowing to monitor the statistics of the parking are made using an intermediate PostgreSQL database that contains all the pre-processed statistics. They are updated each time a new entity is changed in the digital twin. This is done by leveraging the notification mechanism specified in NGSI-LD and implemented in the Stellio context broker.

**Data Quality Assessment**

Data quality check is a valuable and indispensable step for subsequent data processing, analysis techniques and consequent statistics that drive political and economic decision-making. Furthermore, certain methods, such as ML techniques, do not operate with data gaps and data outliers bias results.

This is why some data cleaning pre-processing steps are performed to remove duplicates, empty detections (triggered snapshots containing no information), to correct false labels of vehicle directions, necessary to distinguish entries from exits in single camera solutions.

**Data Security, Privacy, and Protection**

Storing license plates and vehicle passages is a security concern and a privacy issue for the possible identification of vehicle owners. Our solution tackles this problem in three ways.

First, it provides an Edge device that is power and network constrained, indicating there is no permanent access to the internet network other than 4G mobile (Fig. 2). Secondly, license plate numbers are anonymized by hashing once the vehicle technical information has been obtained. A hash is a unique and irreversible encryption code. Messages from the cameras are therefore sent by the Edge computer through an encrypted channel, to the Cloud, where hashes are created and then stored as ID values, while the original numbers are forgotten. And thirdly, the camera does not transfer any image over the network. In addition to facilitating the management of people's privacy, it saves network bandwidth and ensures processing capacity even during network outages. This constrains us to continuously process the video streaming of vehicle at the Edge.

## 4   Camera Efficiency

It happens that the cameras may miss vehicle passages or readings of license plate numbers. This is due to several factors, such as among others, meteorological conditions (fog, heavy rain, lens condensation) and/or changes in camera optical conditions (focus, line of sight, failure). As explained in Sect. 4, this generates incomplete Stay entities in the database, which we make use of for estimating camera efficiency.

Camera efficiency is calculated using the formulae:

$$\begin{cases} p = \frac{n_s}{n_o + n_s} \\ N = \frac{n_s}{p^2} \end{cases} \tag{1}$$

where $n_s$ is the number of correctly detected vehicle passages with associated entries or exits, $n_o$ is the number of incomplete vehicle passages with undetected entries, and $N$ is the total number of detected vehicle passages. We assume that detection efficiencies are the same for entries and exits.

Results show that the camera efficiency is at best operating conditions for Perpignan Sud and the Mas Sabole carparks, 89% and 96% respectively. We then correct our statistics to encompass such data loss.

Camera efficiency is computed daily and continuously monitored. As shown in Fig. 4, a progressive loss in camera efficiency can be observed when the camera is gradually dropping due to improper fixation at the pole, and/or as a precursor to a sudden camera failure due to, for example, a lack of solar energy supply to the battery.

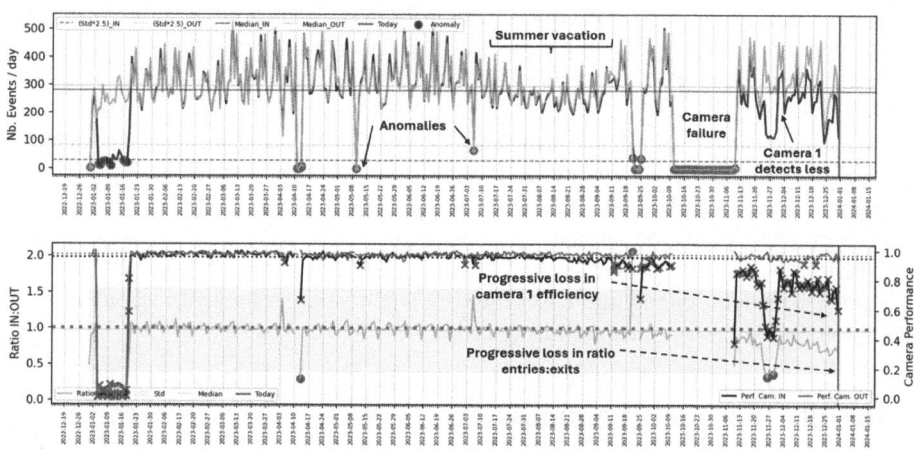

**Fig. 4.** Monitoring KPIs: daily numbers of vehicle entries and exits, camera efficiencies, and ratios of entries:exits. In this example, representing year 2023, one can notice several sudden camera failures, anomalies, impact of school holiday, and a progressive drop in camera efficiency.

# 5   Monitoring

Real-time continuous monitoring of data flows ensures the proper operation of equipment (cameras, MQTT servers). It is performed using data analytics on KPIs and variables, which are then displayed on a Superset Apache or Grafana dashboard (Fig. 2).

The KPIs and variables monitored daily are the number of unidentifiable license plate numbers, the flow and ratio of vehicle entries and exits, and the camera efficiency. The equipment's complete failure is reflected by a sudden absence of data and the operator is immediately notified by email at midnight of the following day or, in the case of the MQTT, if it lasts over 1 h.

Camera failure requires human intervention and indulges costs, whereas a declining camera efficiency detection, using daily monitoring, is the most useful and important indicator to predict camera failure and minimize loss of data. The equipment can then be cured most times by remote control (e.g. lens refocusing), minimized intervention actions and maintenance costs.

## 5.1   Anomaly Detection and Alerting

An anomaly is defined here as absence of data or outlier data points. We experimented with two types of anomaly detection methods: (a) adaptive long-term threshold, and (b) Machine learning methods.

In (a), the threshold value is the standard deviation relative to the long-term median, updated every day. When the camera efficiency falls by 5% or greater, or, when the daily vehicle flow or ratio (entries/exits) falls below 2.5 or 2 standard deviations, respectively, an alert is issued immediately (by email). These threshold values are adaptive since they are updated every day relative to the median of the whole dataset. This method and the values have been fine-tuned and validated over months of feedback experience, to bring best reactivity and maintenance results.

In (b), we wished to test the performance of three ML algorithms: incremental adaptive algorithm for streaming data using River Python toolbox (such as HalfTreeSpace), Sarima and Facebook Prophet. In all three cases, our data reflects too much variability to be properly predicted, although the general weekday behavior could be inferred more-or-less.

For these reasons, our solution uses the simpler and more reliable threshold anomaly detection method (a) for our monitoring task.

# 6   Results and Interpretations

We examined weekday and time frequencies, durations of vehicle passages, types of vehicles and engine fuels, and the "faithfulness" of users. Duration of vehicle passages and attendances relative to weekdays are important indicators of the carpark usages (professional vs. personal). We noticed that all three pilot sites exhibit a similar pattern of three dominant categories of users, validated by a 2 day-long in-situ survey conducted by an independent body.

## 6.1 Overall Frequentation

In 2023, at the Perpignan Sud car park, a total of 105 004 vehicle passages were detected, averaging 288 per day (Fig. 5A). In January, April, October and Novembre, a camera failure explains the data gaps. A clear discrepancy can be observed between weekdays and weekends, whereby weekends are marked by a high vehicle flow on Fridays and Sundays, suggesting a strong influence for weekend travels.

At the Mas Sabole carpark, a total of 24 380 vehicle passages were detected, averaging 67 per day (Fig. 5B). In November, a camera failure explains the data gap, and it was followed in December by a period of gradual drop in camera efficiency after its reservicing, due to inadequate repositioning of the camera in the direction of the target. We noticed that Thursday is the most frequented day, followed by Friday and Sunday.

A common behavioral trend observed at both car parks is the two dominant times for vehicle entries at 7–8 a.m. and 5–6 p.m., and one for exits around 6 p.m., explained by vehicles entering the car park in the morning and exiting at the end of the day. Also, at night-time, traffic in the car park is strongly diminished, as is during the school holidays, particularly during July and August.

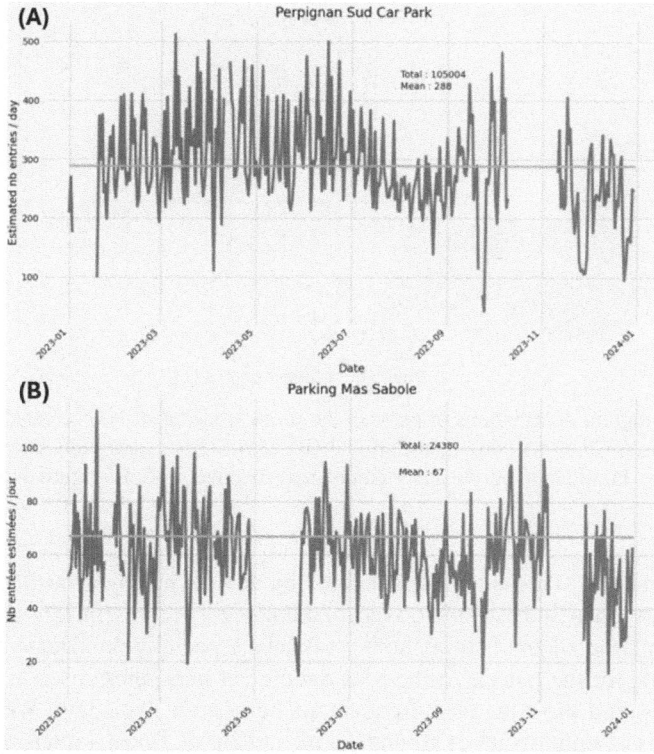

**Fig. 5.** Vehicle flow on the (A) Perpignan Sud and (B) Mas Sabole carpooling carparks during 2023.

## 6.2 Three Main Categories of Users

Based on our statistical analysis of the vehicle passage frequencies, durations, and times, we recognized three main categories of users, numbered on Fig. 6 and Fig. 7, and described below.

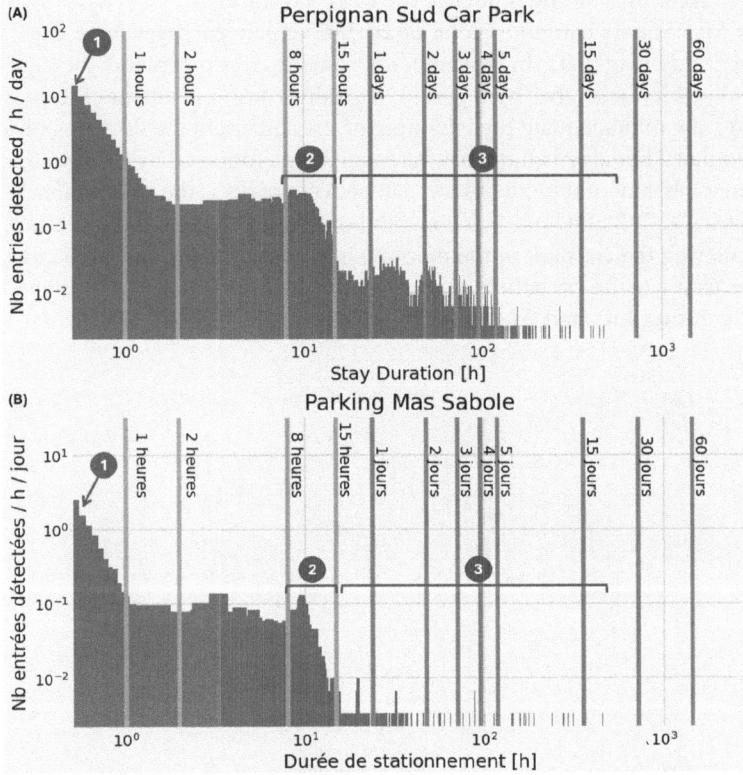

**Fig. 6.** 1D-Histogram distributions of passage durations (log-log scale) corrected for efficiency and breakdowns for the (A) Perpignan Sud and (B) Mas Sabole carpooling carparks, showing hourly and daily durations in purple and green bars, respectively. (Color figure online)

**Driver Carpoolers.** This category is defined by vehicle passages lasting less than 10 min. At the Perpignan Sud car park, drivers of this category represent 75% of all passages, and show regular attendance throughout weekdays especially on Fridays and Sundays (primarily exits for the latter). At the Mas Sabole car park, these users represent 25% of all passages, and show the most frequent attendance on Thursdays. We will see that this category is essentially active around 7 a.m. and 5 p.m. (local wintertime) which we interpret as carpooling drivers that pick-up and drop-off passengers in the mornings and evenings, respectively.

**Passenger Carpoolers.** This category is defined by vehicle passages lasting between 5

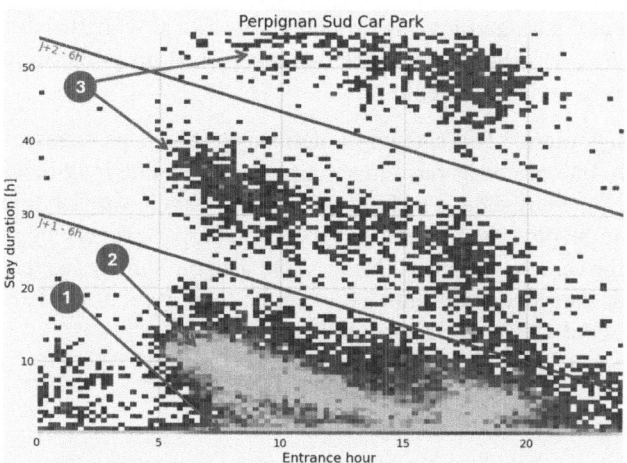

**Fig. 7.** 2D-Histogram distributions of passage times and durations (log-log scale) corrected for efficiency and breakdowns for the Perpignan Sud carpooling carpark. The distribution for the other car park is similar.

and 15 h (essentially during daytime). Activity is highest during weekdays, particularly on Thursdays for both car parks. They represent ~10% and 5% of the total users of the Perpignan Sud and the Mas Sabole car parks, respectively. We will see that they are

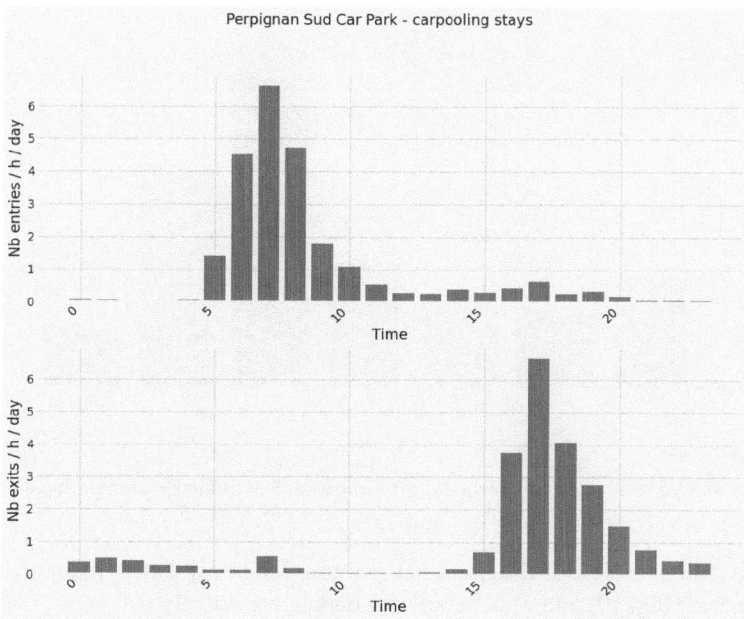

**Fig. 8.** Daytime attendances for the passenger carpoolers at the Perpignan Sud carpark, (top) entries, (bottom) exits.

mostly arriving at 7 a.m. and leaving around 5 p.m. (local wintertime) (Fig. 8). These are interpreted as carpooling passengers that are picked-up in the morning by driver carpoolers and returned early evening.

**Long-Term Carpoolers.** This category is defined by vehicle passages lasting at least 15 h and up to 30 h. Interpreted as passenger carpoolers traveling long-distances generally for 2 to 3 days. Whereby data is lacking for the Mas Sabole carpark to properly assess this category, on the other hand, the Perpignan Sud carpark is very popular on Fridays and Saturdays for vehicle entrances (Fig. 9), and also on Sundays as they return to exit the carpark. This category represents ~10% of the total users of the Perpignan Sud car park, and only about 1% for the Mas Sabole car park.

### 6.3  Usage Patterns

We noticed a drop in the attendance of both car parks during school holidays, particularly for the passenger carpoolers at the Perpignan Sud car park, and for long-term carpoolers at the Mas Sabole car park.

At the Perpignan Sud car park, a dominant professional usage is reflected by the significant proportion of passenger carpoolers, the latter's high weekday attendance, and the clear school holidays impact. This agrees well with its location on the outskirts of a major city, where city workers are known to commute from rural habitations. On the other hand, the Mas Sabole car park is in a more rural setting and is visited mainly for personal and/or touristic reasons, as the survey taught us.

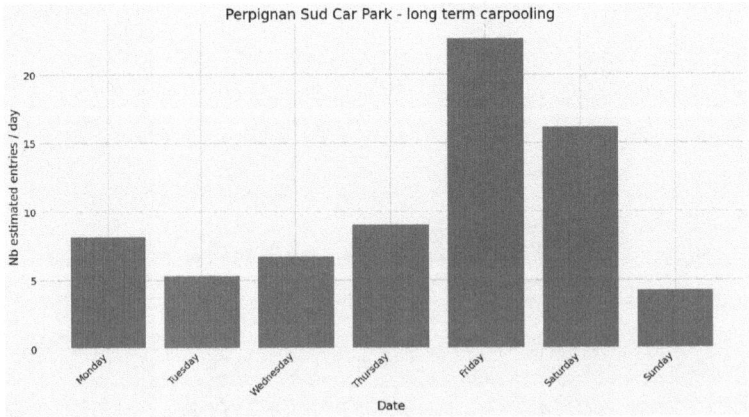

**Fig. 9.** Weekday attendance for the long-term carpoolers at the Perpignan Sud carpark.

About 75% and 60% of the users never return, and only ~10% regularly visit the Perpignan Sud (Fig. 10) and Mas Sabole car parks, respectively.

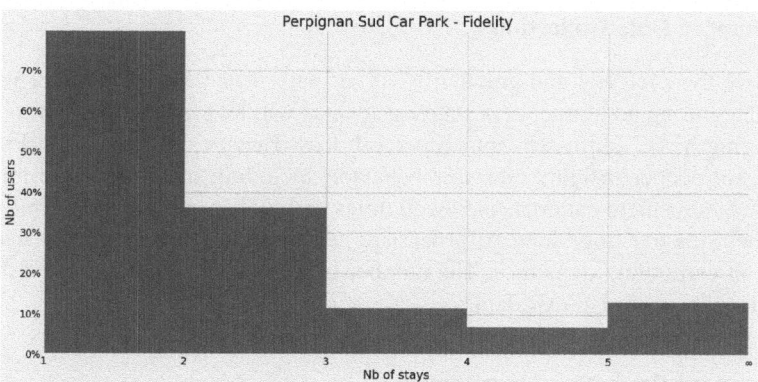

**Fig. 10.** Frequency of returns of "Stays" entities (entrances are associated with an exits) at the Perpignan Sud carpark over the year 2023.

### 6.4 Types of Vehicles

Understanding the types of vehicles that participate in carpooling actions is interesting to understand user habits, and for instance to adequately furnish the car parks with electrical recharge requirements. Most vehicles (>93%) are personal vehicles with combustion engines. Attendance of hybrid or totally electric cars at the Perpignan Sud car park represents ~10% of all carpooling drivers and is twice as high for this category of users than the others. A very small proportion of electric cars remain stationed for several days at a time, whose presence at all is unexpected. Both car parks contain 2 electric car recharge emplacements.

## 7  Assessment of $CO_2$ Emissions Avoided Thanks to Carpooling

The carbon impact in this case is based on the amount $CO_2$ emissions avoided into the atmosphere, thanks to carpooling practices. It is estimated by summing the $CO_2$ emissions per kilometer for a given time range (here 1 year) for each passenger carpooler's vehicle parked for over 5 h, multiplied by the roundtrip distance the passenger undertakes with a carpooling vehicle.

This distance is defined in our calculations differently for both carparks, since they portray different usages (see Sect. 6.3) and are established by the complementary survey results founded from passenger carpooler answers: 143 km and 47 km for the Perpignan Sud and Mas Sabole carparks, respectively. Should these distances considered evolve over time, this parameter is made easily configurable in the dashboard.

Our computations indicate that an estimated 277 tons and 16 tons of $CO_2$ emissions were avoided in the atmosphere over the year 2023 for the Perpignan Sud and for the Mas Sabole car parks, respectively.

## 8  Future Work

Building upon these results, several areas for future research and development have been identified to further enhance the monitoring and analysis of carpooling car parks.

## 8.1 Enhanced Data Collection

To improve the accuracy and granularity of car park usage analysis, future work will involve integrating additional data sources, such as wireless traces from users' devices. By capturing MAC addresses from Bluetooth Low Energy (BLE) and Wi-Fi signals, we can gain deeper insights into user behavior, including tracking of the number of passengers, movement patterns, and dwell times within the car park areas. This data can complement the existing video AI processing, enabling a more comprehensive understanding of carpooling dynamics. The number of users may also be inferred from mm wave radar sensors or LIDAR data.

## 8.2 Advanced Predictive Algorithms

Developing predictive algorithms that leverage the collected data to forecast car park usage patterns will be a key focus. The methods of [1, 2] would be a good starting point for this. These models could predict peak times, anticipate future demand, and suggest optimal car park management strategies. By utilizing machine learning techniques, these forecasts could be continuously refined and adapted based on real-time data, improving their accuracy and utility for decision-makers.

## 8.3 Automated Alerting Systems

Another crucial area of future work is the development of a sophisticated alerting system. This system will automatically notify authorities or users of anomalies or important events, such as overcrowding, or unusual stay durations. These alerts could be triggered by pre-defined thresholds based on Key Performance Indicators (KPIs) or by anomaly detection algorithms that identify unusual patterns in the data.

## 8.4 Integration with Broader Smart City Initiatives

To maximize the impact of our IoT solution, future efforts will focus on integrating the carpooling monitoring system with broader smart city infrastructures. This includes linking with public transportation systems, traffic management centers, and environmental monitoring networks to create a holistic view of urban mobility and its environmental impact. Such integration could enable more coordinated and effective urban planning and management.

## 8.5 User Experience and Privacy Enhancements

Future work will also involve gathering user feedback to refine the system's functionality and user interface, ensuring it meets the needs of both end-users and local authorities. Additionally, as new data collection methods are introduced, robust privacy safeguards will be implemented to protect users' personal information, ensuring that the system complies with relevant data protection regulations.

### 8.6 Scalability and Long-Term Sustainability

The final area for future work is the scalability of the solution. We plan to explore the deployment of this system across a larger network of car parks, including different geographic regions and urban settings, to assess its adaptability and long-term sustainability. This expansion will also provide valuable data to refine our algorithms and system architecture further, ensuring the solution can support a wide range of use cases and environments.

## 9  Conclusions

This study presents our solution for carpooling car parks based on real-time monitoring. We aimed at a better understanding of carpooling users' behavior and to estimate the carbon impact for the purpose of such car parks. Our solution is composed of both an Edge and a Cloud component whereby data is collected by a camera hosting algorithms of image processing and encryption processes, asserting data privacy, security, and protection. The anonymized data is stored, cleaned, augmented, analyzed, and monitored from the Cloud architecture.

Our solution was implemented on three carpooling carparks, near Perpignan City in Southwestern France, for which two are still in operation today. Our data analysis for the year 2023 shows, for most pragmatic monitoring, the best KPIs to use are not only the daily number of vehicle entries and exits and the ratio between entries and exits, but mostly the camera's efficiency.

We were able to observe three dominant categories of users for both carparks: (i) carpooling drivers, whose passages last less than 10 min, (ii) carpooling passengers, who park their vehicle between 5–15 h at daytime, and (iii) long-term carpooling passengers, who generally station from 1 to 3 days.

Our results show two different usages for these carparks: professional and private/touristic and were validated by onsite 2 day-long surveys at carpark, helping us establish the approximate distances travelled by the passenger carpoolers for our computations of the estimated $CO_2$ emissions avoided to the atmosphere during that year.

**Acknowledgments.** This work was financed by the Conseil Départemental des Pyrénées Orientales, France.

## References

1. Sutjarittham, T., et al.: Measuring and modeling car park usage: lessons learned from a campus field-trial. In: 2019 IEEE 20th International Symposium on "A World of Wireless, Mobile and Multimedia Networks" (WoWMoM). IEEE (2019)
2. Sutjarittham, T., et al.: Monetizing parking IoT data via demand prediction and optimal space sharing. IEEE Internet Things J. **9**(8), 5629–5644 (2020)

3. Orazi, G., Abid, A., Le Gall, F., Richard, A., Martzluff, Y.: Parking monitoring and usage statistics using an Edge-Cloud solution based on FIWARE. Presented at the 14th ITS European Congress, Toulouse, France, June 2022
4. Abid, A., Lee, J., Le Gall, F., Song, J.: Toward mapping an NGSI-LD context model on RDF graph approaches: a comparison study. Sensors **22**(13), 4798 (2022). https://doi.org/10.3390/s22134798

# Navigating the International Data Space To Build Edge-Driven Cross-Domain Dataspace Ecosystem

Parwinder Singh[1]([✉])[iD], Nirvana Meratnia[2][iD], Michail J. Beliatis[1][iD],
and Mirko Presser[1][iD]

[1] Department of Business Development and Technology, Aarhus University,
7400 Herning, Denmark
{parwinder,mibel,mirko.presser}@btech.au.dk
[2] Eindhoven University of Technology, 5612 AZ Eindhoven, Netherlands
n.meratnia@tue.nl

**Abstract.** Data is an asset for the modern industrial era that plays a crucial role in driving intelligence through machine learning techniques and methods for business, technical, economic, and social decision making. Generally, data generated in a specific domain is often used in other domains, which requires cross-domain data integration (CDDI) and sharing across edges. Here, an edge represents an organization, system, or entity with computing data processing capabilities. This CDDI and sharing across edges need data sovereignty to govern the relevant usage context, wherein the sharing edge domain's data context must be respected by the shared edge domain. This edge domain's data context can be seen as a medium to define rules for associated data usage, access, and identity management. This can be achieved by the common dataspace vision of the European Union for CDDI to turn the European market into a unified data-driven European market. International Data Space (IDS) is such an effort backed up by the European Commission to develop relevant standards and specifications for CDDI. However, even though a vast amount of IDS ecosystem information is available online, it is scattered and hard to navigate. This obstructs the use and adoption of the IDS at a desired pace and in a simplified manner. This study contributes to this need by converging the information in one place and leveraging theoretical and pragmatic insights on building IDS-based edge-driven dataspace in real-world scenarios. We dive into a wind industry supply-chain-specific use case realized through a locally developed IDS platform to showcase and validate the use of CDDI.

**Keywords:** Dataspace · International Data Space (IDS) ·
Cross-domain · Data Integration · Data Sovereignty

## 1 Introduction

Data has recently become a critical asset for modern industries worldwide. This can also be seen in the fourth industrial revolution, i.e., Industry 4.0, which

M. Presser et al. (Eds.): GIECS 2024, CCIS 2328, pp. 151–168, 2025.
https://doi.org/10.1007/978-3-031-78572-6_10

revolves around Big data and associated technologies such as Cloud Computing, Artificial Intelligence (AI), Machine Learning (ML), and Internet of Things (IoT) [1] for the processing of data with 4V (Volume Veracity, Velocity, and Variety) characteristics [2]. Data is crucial in deducing intelligence by applying ML-based models, including regression, classification, clustering, association, and control methods [3]. Applying these methods to data generates informational insights that help business, technical, economic, and social decision making [4]. Therefore, data has an economic value that can play a significant role in the sustainability and growth of any given industry or domain [5]. In addition, with the advancement of 5G/6G networks and Internet of Things (IoT) technologies, the generation and collection of data has become more viable [6].

Generally, data generated in a specific organization is often associated with its domain knowledge or context [7,8]. In the current study context, an edge (following edge computing paradigm) represents an organization, system, device, or any entity that has significant infrastructural and system capabilities for different data processing requirements under Dataspace (DS) ecosystem [9,10]. Therefore, each edge will have an associated domain context, which can be represented by the ontological methods that define how different entities within domain w.r.t data are linked to each other [7,8]. This domain-level informational context can also include the terms or conditions under which the relevant data of this domain can be used in other domains [11]. We call this contextual use of data, which is a fundamental block in achieving data sovereignty [12]. Most of the time, data generated within a domain remains in the domain boundaries, making data silos and failures to achieve the mutual gain (e.g., generating revenue streams or services) that can arise from its sharing with others [9]. This happens for multiple reasons, including domain-specific problems, competitiveness feeling, and lack of data-sharing methods that support data owners' interests, trust, security, transparency, privacy, complexity, interoperability and integration, and lack of knowledge [10]. All these result in data not being used to its full potential, which otherwise could have created numerous opportunities [9]. This problem can be addressed through the concept of Dataspace. Dataspace is not a new concept, but its definition and usage have evolved. However, one thing that remains persistent w.r.t its definition is that *data owners always perceived to have complete control over their data* [13], which also reflects through data sovereignty objective [14]. Authors of [14] defined Dataspace as a *"federated, open infrastructure for sovereign data sharing based on common policies, rules, and standards"*. From EU perspective [14,15], the Dataspace is supposed to have the following functional characteristics:

- **Security and Privacy Preserving Infrastructure**: The Dataspace should have a robust IT infrastructure (generally distributed in nature) that supports the security and privacy of data. This infrastructure should enable data pooling, accessing, processing, and sharing while safeguarding individual's or sharing domain's privacy rights.
- **Data Governance Methods:** Dataspace comprises of administrative and contractual rules or policies that will govern different rights to different data

operations such as access, process, use, share, etc. It is important for exchanging clear information in the form of guidelines and responsibilities for data management while data is crossing its organizational boundaries.

- **Protecting Data Owner Rights:** Data holders always have control over their data life cycle regarding its usage conditions, access, and permissions related to data management to define data usage contexts and intentions while data is being shared in cross domains [9]. From the data owner's perspective, it is important to determine how the data is being used or accessed to protect the data owner's rights over data.
- **Monetary Value:** Data made available in a Dataspace, based on a volunteer basis, can be used against remuneration or without cost, depending on the decision of the data owner.
- **Open Participation:** The Dataspace should support open participation for all, including any organization or individual, adhering to participation rules, ensuring a fair and level playing field by avoiding any discrimination for any participant.

To support cross-domain data integration (CDDI), the EU has taken many initiatives to promote data sharing, making it open and accessible to all with fair usage [16]. The initiative includes developing the European Data Strategy [17], Common European Data Spaces [13], for which a working document was released in 2022, a cross-sectoral legislative framework comprised of the Data Governance Act [18], Data Act [19], Implementing Act on High-Value Datasets also known as Open Data Directive [20]. EU also sets DSSC (Data Spaces Support Center) to coordinate and govern the actions of these initiatives. In addition, the EU has also set up a European Data Innovation Board, which is the consultative and advisory body that provides guidelines for the interoperability of Common European Data Spaces [14].

The objective of all aforementioned initiatives is to enable the opening up and sharing of data in different domains through CDDI, so that the true potential of converged data can be exploited in different dimensions. Figure 1 illustrates the CDDI concept among different edge domains, which can generate cross-organizational value chain streams [10]. However, this requires standardized methods for harmonizing data across disparate edge domains [9]. In addition, equally important is safeguarding the interests of data owners through implementation of data sovereignty. This entails respecting the contextual and regulatory frameworks governing the data life cycle within each domain during the CDDI or sharing process [18]. Essentially, CDDI necessitates technical integration and adherence to legal and regulatory standards to ensure integrity and security of data while facilitating effective cross-domain collaboration. In CDDI, the domain-specific data context can be seen as the boundary at the edge where the data owner can define usage, access, and identity control or management of associated data operations. European Union foresaw building such data spaces for CDDI, which is essential to support the vision of turning the European market into a unified market based on data-driven value streams [13].

Aligning to the EU vision, an effort for International/Industrial Data Space (IDS) was initiated by the Fraunhofer Institute in 2015 [21] and backed up by the European Commission to develop relevant standards and specifications. To this end, IDS Association was formed in 2017 and developed IDS Reference Architecture Model, Dataspace protocol specification, and related IDS prototype, which can uptake the CDDI activity aligning EU vision to converge data among European industries. However, understanding the IDS ecosystem and setting it up at the local or regional level is a huge challenge due to the vast amount of scattered information available online, which is complex and hard to navigate. This obstructs the use and adoption of IDS in a simplified and understandable manner. This raises an important research question (RQ), namely,

*How can IDS-based standards, specifications, and reference implementation be used to build a local or regional Dataspace for a target use case focusing on edge-driven cross-domain data integration?*

This paper contributes to addressing the above RQ by converging the information around the EU's vision of building Dataspaces, related work, and IDS specifically in one place and in a simplified manner. We leverage theoretical and pragmatic insights on building IDS-based Dataspaces in real-world scenarios through a wind turbine industry supply chain-specific use case to achieve CDDI. In addition, we also provide knowledge on how a specific use case can be semantically modeled through the IDS information model, followed by its implementation through a locally developed IDS platform prototype, which supports the requirement for CDDI with functional compliance and validation.

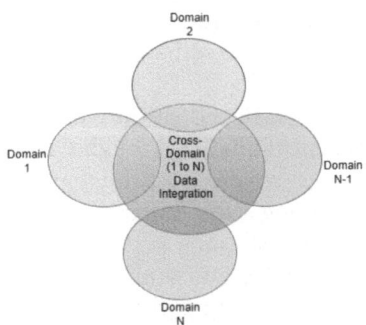

**Fig. 1.** Cross-Domain Data Driven Dataspace

## 2    Literature Review

Applications of Dataspace in diverse use cases can bring various advantages for CDDI, such as the wind industry supply chain or the role of digital transformation in roll-to-roll printing in Industry 4.0 settings [22,23]. However, CDDI must address interoperability and heterogeneity challenges (through meta-standards) induced by diverse standards and protocols used in different organizational boundaries for data sharing, reuse, and distributed operations [9]. [24] presents the role of Blockchain in the IDS-based data exchange ecosystem for data sovereignty and implementation of a clear housing data trading platform. [25] has presented in-depth knowledge on security aspects for usage and access control for achieving data sovereignty, identity, and integrity management of components to embed trust through Public Key Infrastructure (PKI) methods among different participants in the Dataspace ecosystem. [26] presents an architectural framework to develop a cross-domain distributed infrastructural ecosystem based on Processing, Service, and Data context by extending the traditional Cloud-Edge-Device (CED) continuum by introducing three new layers, namely, Semantic, Convergence, and Dataspace integration.

[27] introduced a prototype for IDS Security Components tailored to the Textile and Clothing Industry, ensuring data sovereignty and facilitating secure interactions among supply chain participants. [28] suggests leveraging Dataspaces and digital twins to implement the Industry 4.0 vision, exploring key components such as integrating digital twins within Dataspaces according to the latest Industry 4.0 Asset Administration Shell specifications. [29] targets bridging the gap between IDS and Plattform Industrie 4.0 (PI4.0) by analyzing their concepts and tools and recommending a generic approach for combining different technologies, irrespective of specific IDS or Industry 4.0 implementations. [30] proposed the establishment of an International Testbed for Dataspace Technology, i.e., a testbed for developing and testing data platform technologies' interoperability, portability, and customizability. [31] presents the reference implementation of IDS specifications, named *True Connector*, developed by Engineering. Authors of [32] present the learning from implementing IDS-based Dataspace Connector (DSC) and their interactions at a high level. This is quite similar to our work, but it only covers the architectural perspectives. In a way, we are extending their study with the latest findings and insights for IDS implementation.

## 3   Methodology

The constructive research approach [33] has been followed in our study to guide the design, development, and deployment of our IDS-based Dataspace prototype, ensuring its effective and sustainable adoption. We studied many EU Commission policies and working documents to determine the role of Dataspace in the futuristic industrial landscape in terms of CDDI. Finally, we recorded theoretical information, such as relevant background and semantic modeling of IDS from a target use case perspective, as the basis for our study to conduct practical experiments and to build an IDS-based Dataspace ecosystem. In this context, we have investigated the IDS Connector, IDS Testbed [30], and True Connector (from Engineering) [31] as IDS reference implementation [32]. After that, based on this open-source implementation [31,34], we developed our prototype with default settings for IDS-based data integration in different domains and participants acting as data producers and consumers. After completing the prototype with default implementation, we applied our wind turbine supply chain use case to an IDS information model. We then prototyped and evaluated it, focusing on IDS connector implementation, standards compliance, and protocol message exchange sequence. The relevant services of IDS implementation have been deployed following the Microservices design architecture. For this, Docker-based toolchain was used over virtual infrastructure to instantiate and validate the prototype. An online Resource Description Framework (RDF) graphic visual tool was used for semantic modeling (ontology) visualization. During prototype development, many challenges, such as the deployment of services failure, semantic modeling adaptation, API validation failures, inter-component connectivity, and service discovery-related issues, were experienced and solved, and the relevant lessons learned have been documented.

As a contribution, we have advanced the relevant knowledge for building IDS-based Dataspace at edge, focusing on the target use case from theoretical and practical perspectives. We believe this will also contribute to accelerating the adoption of IDS-based standardized Dataspace implementation across Europe, specifically focusing on the industrial region of Denmark in our study.

## 4    Use Case Specific IDS Development

### 4.1    IDS Background

IDS offers a framework to exchange data in a cross-domain environment offering data sovereignty, including access and usage control [35]. The Reference Architecture Model [35] of IDS consists of the following layers:

- *Business Layer:* provides different roles (data providers, consumers, brokers, etc.) and their interactions, encompassing contracts and relevant data usage policies.
- *Functional Layer:* provides requirements for trust, security, data sovereignty, interoperability, value-added applications, and data marketization.
- *Process Layer:* specifies the processes and interactions between components for data exchange, including policy enforcement.
- *Information Model:* provides a standardized way to describe and represent data entities using the RDF/JSON-LD contextual format and IDS ontology, enabling interoperability.
- *System Layer:* defines and realizes the technical components, unified namespace, and their interactions for implementing the IDS infrastructure.

IDS Connector is a key component in realizing IDS architecture that facilitates secure, traceable, and trusted data exchange among stakeholders or participants in the cross-domain data ecosystem [21]. The connector may have multiple components, such as data adapters, security/identity modules, policy/rule engines, metadata repositories, and cross-domain communication interfaces, to support IDS functionality effectively [32]. This connector is a main gateway for data exchange between participating parties, including data providers, consumers, and intermediaries. It enables the organizational edges to share, reuse, and integrate data within the IDS-distributed infrastructure, including deployment over cloud, edge, on-premises, etc., depending on the specific requirements of the target use case. This connector enforces data governance to achieve data sovereignty through policies and access controls, ensuring data is accessed and shared according to predefined rules, regulations, and exchange protocols, providing interoperability among participating entities. This includes security mechanisms such as authorization, authentication, encryption, and access control to protect CIA (confidentiality, integrity, and availability) Triad. It also offers the publication of catalogs through a metadata broker that provides the availability of metadata to the requesting participant for the data to be exchanged [31]. This metadata context provides semantics and provenance information to facilitate data discovery, understanding, and utilization. IDS Connector can be applied to various use cases across industries such as healthcare, smart cities, logistics, energy, etc.

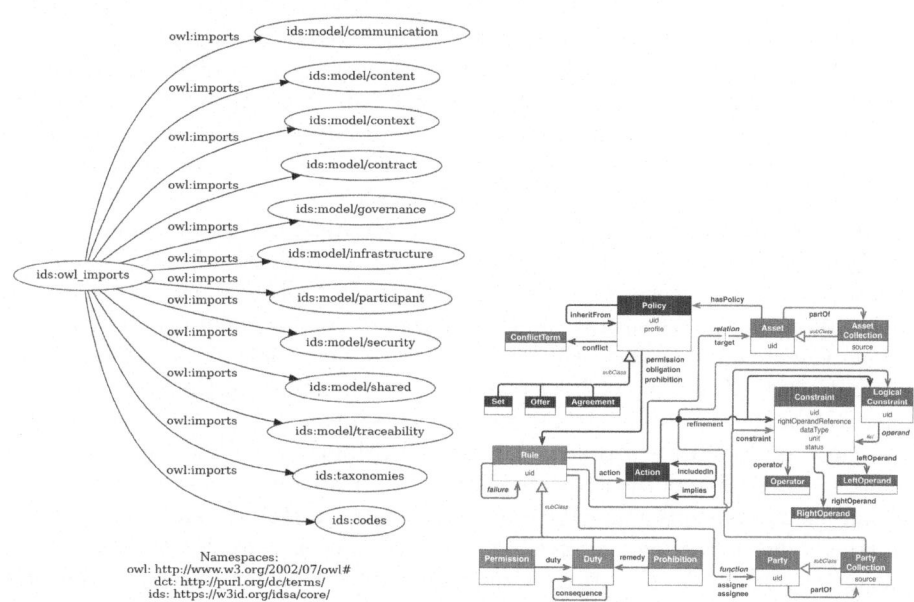

(a) IDS Information Model (Code View).     (b) ODRL Model. Sourced from [36].

**Fig. 2.** IDS and ODRL Information Model

## 4.2   IDS Information Model

IDS information model is the key core element for the participating entity to participate in an IDS-based Dataspace [11]. The IDS information model defines metadata or data elements using a common vocabulary, technically called ontology. It must be used for data modeling while sharing the data in cross-domain integration. The relevant information model is shown in Fig. 2a. IDS semantic information model bridges the IDS ecosystem's expression, infrastructure, and enforcement components for delivering seamless data integration and interoperability in cross-domain integration [11].

IDS info model uses the Open Digital Rights Language (ODRL) model, shown in Fig. 2b, [36] to define the relevant usage policies and permissions to achieve data sovereignty. ODRL, as shown in Fig. 2b, provides a standardized vocabulary and information model for representing digital rights expressions, such as permissions, prohibitions, obligations, and constraints associated with digital content or services. It allows data holders to specify how their data can be shared, distributed, or consumed by other participants of the IDS ecosystem.

Figure 2a provides a high-level view of IDS information model/ontology generated based on IDS source code [34]. It is observed that IDS ontology (OWL-based) imports from different models related to Communication, Content, Context, Contract, Governance, Infrastructure, Participant, Security, Shared, Traceability, Taxonomies, and Codes. Each importing model/ontology encapsulates specific functional entities relevant to its domain. For instance, Communication encompasses components like AppRoute, Endpoint, Message, and Proxy, reflecting elements involved in data transmission. Context incorporates SpatialEntity and TemporalEntity, which are indicative of spatial and temporal considerations within the system's environment. Content includes Artifact, Asset, DataApp, DigitalContent, Language, MediaType, PaymentModalities, Representation, Resource, UsageControlObject related ontologies suggesting diverse content types managed within the IDS system. Contract includes ODRL-based Action, Constraints, Rules, BinaryOperator, UsageControl, Contracts, Left-Operand, and UsagePolicyClasses, which are elements pertinent to contractual agreements. Similarly, Infrastructure encompasses infrastructure components such as AppStore, Broker, Catalog, ClearingHouse, ConfigurationModel, Connector, DAPS, IdentityProvider, ParIS, and PublicKey, hinting at the foundational elements supporting IDS soft infrastructure [15] system operations. In a nutshell, this info model represents the IDS system's architecture and organization at the implementation level, facilitating a better understanding and communication of its intricacies.

## 4.3   Use Case

Our use case, presented in [22], and [10], revolves around the supply chain events associated with wind turbine operations, with a particular focus on full product life cycle and bolt-related activities. This illustrates the collaboration among various stakeholders across different domains. The bolt vendors or suppliers are at the heart of this supply chain, responsible for manufacturing bolts in their production units and delivering them in batches to the turbine operators. The turbine operators, who own and manage the turbines, rely on third-party service engineers for maintenance and operational support. Additionally, an insurance

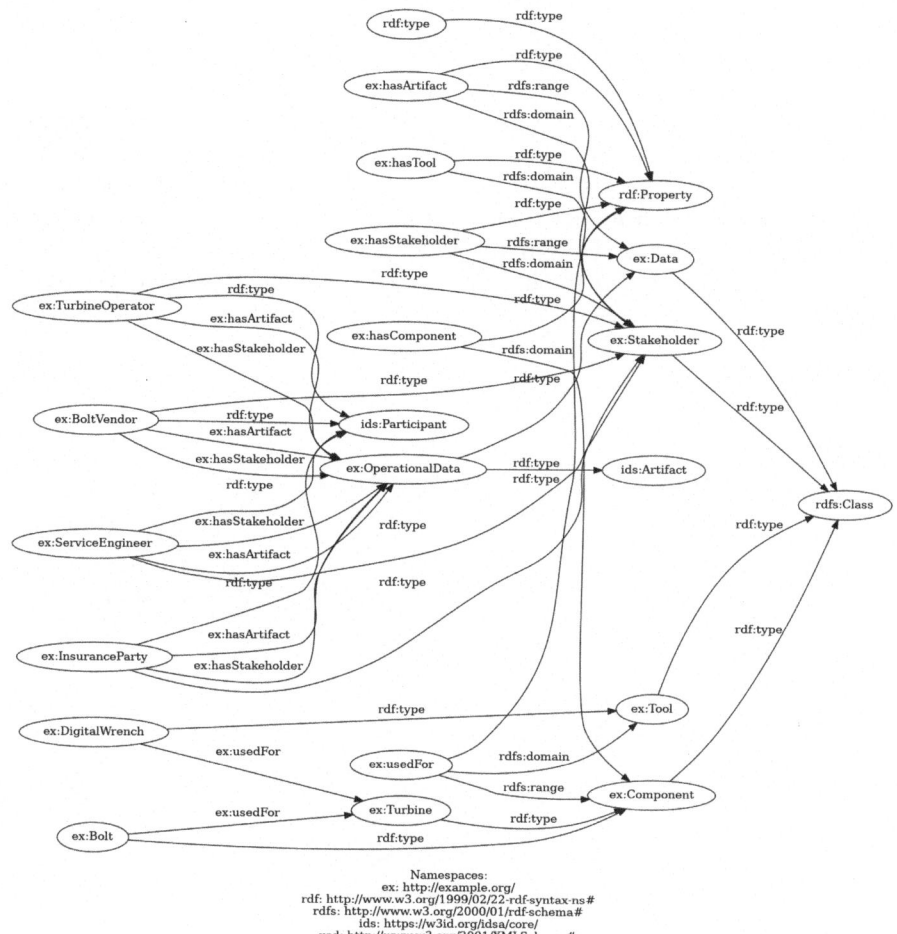

**Fig. 3.** IDS Based Wind Turbine Bolt Operations Info Model

partner plays a crucial role in mitigating risks by offering insurance services to protect the interests of the turbine operators against unforeseen events. Wind turbines have multiple components, including blades, towers, bolts, motors, electrical cables, rotors, electronic sensors, etc. However, our focus is on bolt-related operations management within the turbine context. Bolts play a critical role in turbine operations, and ensuring their timely maintenance is paramount. The main objective of this use case is to facilitate the exchange of bolt-specific operational information among various stakeholders, including turbine operators, bolt suppliers, service engineers, and insurance partners across different domains. This relationship is illustrated in the semantic modeling diagram of Fig. 3. This shows the semantic relationship among stakeholders operating in different domains, tools, and components like bolt-batch and turbine. Operational data represents the event data the Service Engineer recorded for the bolt operation over the turbine. Then, this data is exchanged for different purposes or actions and provided to different stakeholders, such as bolt suppliers and insurance parties. For example, the insurance party can validate the quality of the bolt operation during the assembly of the turbine to settle any claim in the future. Figure 3 shows that the stakeholders will be treated as Participants as per the IDS information model. This information model is developed using the ontology provided by the IDS GitHub repository [34] and the use case-specific data model [10, 22] mapped to this IDS model to get it processed through the IDS connector running over the IDS prototype explained in the next section.

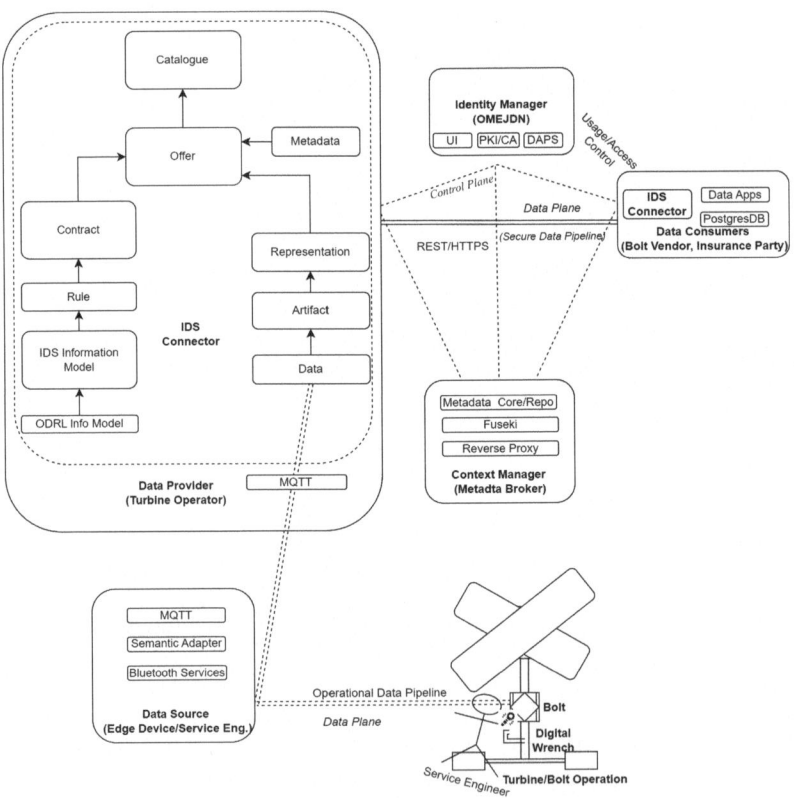

**Fig. 4.** IDS Provisioned Use Case Specific System Architecture

### 4.4   System Context

Fig. 4 provides the system architecture for our use case presented earlier in Sect. 4.3. There are multiple architectural blocks whose description is given below:

– **Data Source**: is the turbine site, where the physical operation on the bolt is performed. The Service Engineer uses the digital wrench to perform the bolt-tightening operation and scan the turbine and bolt-related bar codes as their digital identity. The relevant operational events are transmitted as data streams through a data pipeline from these physical objects to a nearby edge computing device. This data source, consisting of multiple services such as Bluetooth (BLE) protocol stack termination, MQTT agent, and Semantic Adapter (service to transform the BLE data into JSON format) are running at the edge, technically a Raspberry Pi device in our prototype setup, and sending the data towards the Data Producer.

– **Data Producer**: is an organizational edge [9,10], which has on-premises setup, consumes data over the MQTT interface from Data Source and forwards the data towards the IDS connector that is also running as a service on the same system. This is called Data Producer from the IDS ecosystem view, as the data stream is offered/produced as a catalog in the Dataspace.

– **Data Consumers**: are other organizational edges [9,10], which have on-premises setup and consume data over the IDS interface (HTTPS/REST in our case) from the catalog made available to Dataspace by the Data Producer. Multiple applications can run on the consumer side that needs the consumed data to offer innovative services. In our case, consumers such as bolt vendors and insurance partners are using this data to reconcile their entries and monitor the quality of their supplied products (bolt batches) to settle any issues based on received data in the future.

– **Identity Management(IDM)**: is the security module of the IDS ecosystem, which is responsible for establishing trust between the data-exchanging parties or participants. The IDM also consists of a functional component called Dynamic Attribute Provisioning Service (DAPS) within the IDS [32]. This service manages the digital identity of participants/systems in IDS that rely on different attributes linked to that identity [27]. The DAPS service provides dynamic and updated attribute information about participants and connectors. IDM consists of services and mechanisms designed to ensure the CIA-Triad of data exchanged between participants in the IDS ecosystem. It provides the requested parties a DAPS token as per OAUTH2 [37] token workflow and standards. It also uses the Public Key Infrastructure (PKI) to generate X.509 certificates [38]. IDM architectural block also consists of an open-source implementation of OAUTH2, which Fraunhofer provided and called *Omejdn* in the IDS implementation context.

– **Metadata Broker**: is the cataloging system where the offers, resources, or catalogs are registered, discovered, and made available to different participants in the IDS ecosystem [32]. In our case, the Data Producer registers its catalog offerings (with resources data/metadata) with the broker. The data consumer can query the availability of desired/interested resources (data/metadata) from the metadata brokers to fetch the endpoints of the data producer and kick off the data exchange negotiation process. This metadata broker is also responsible for managing metadata-related contexts. In our case, there is a service called *Fuseki* running as part of the IDS context management architectural block, which is an abstract name for Apache Jena Fuseki, i.e., a SPARQL Protocol and RDF Query Language server [39]. This provides services for semantic queries based on RDF standards and ontologies-driven data models. It allows requesting parties within the IDS ecosystem to query certain catalogs or resources based on RDF query strings, which typically follow the Subject-Verb-Predicate (aka triple) structure [39]. This semantic capability enables data producers and consumers to access and retrieve relevant data within the IDS ecosystem efficiently. In addition, there is also a proxy service running, which is at the forefront and exposes the interface for all broker services to the external world through a unified URL interface.

The architectural message flow consists of two communication planes: Control Plane (CP) and Data Plane (DP). CP is responsible for establishing contracts, registering catalogs or artifacts, authentication, authorization, and negotiation between data producers and consumers. On the other hand, DP manages the actual forwarding and processing of data between producer and consumer.

## 4.5   System Validation

The system context explained earlier is deployed independently in two different modes, i.e., (i) in the Azure cloud environment and (ii) in the on-premises environment (on a local laptop). In both cases, the virtual infrastructure was instantiated using virtualization services with the following specifications: VCPU: 2cores; RAM: 8GB; Hard-Disk: 20GB; OS: Ubuntu20.04. A VM was created using these specifications, followed by the necessary Docker and Docker-Compose packages installation for microservices instantiations [40]. After completion of the Docker platform, the Gitlab repository was cloned from [34] to build our use case-specific IDS environment. [34] provides detailed steps to clone, install, and provision the IDS environment and start validation for different IDS standard API requests (based on the Postman tool).

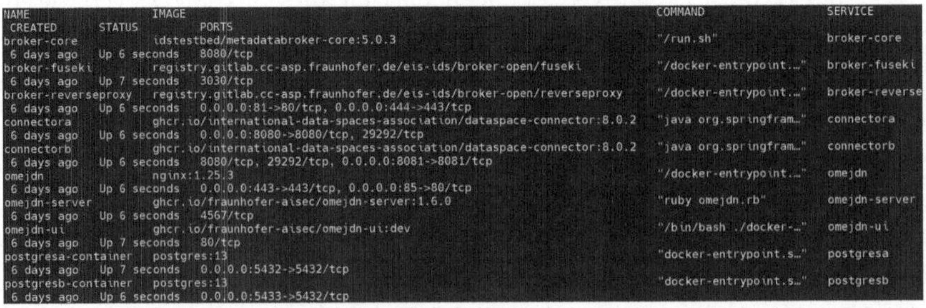

**Fig. 5.** IDS Microservices Based Deployment View

Figure 5 depicts the deployment view of all the microservices running successfully and ready to be validated for handling API requests between the Data

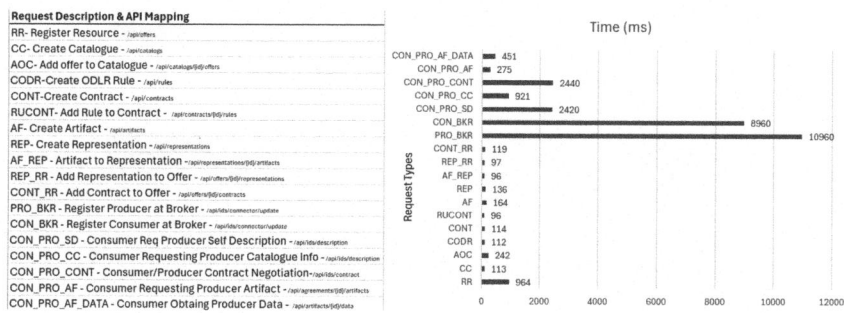

**Fig. 6.** IDS Validation Results - Request Vs Response Time

Producer and Data Consumer. All API requests and their end-to-end response time are shown in Fig. 6. One can see that it is less time-consuming when the requests are landing directly at specific components such as Data Producer or Consumer. However, the response time is significantly increased when the Metadata Broker comes into play for the registration of an entity (producer/consumer) at the metadata broker or the initial discovery of information by an entity in the IDS ecosystem when the broker needs to find the availability of a catalog. This happens because the IDM intercepts the flow when one entity wants to interact with the other party. In the entity registration case, the entity needs to be validated by IDM through the OAUTH2 token workflow before allowing access to the Metadata Broker. Similarly, when a Data Consumer interacts with the Producer directly, it has to be authenticated, authorized by the IDM, as a trusted third party, in order to start negotiation with the producer. This also demonstrates the usage of access control and data sovereignty-related functional validation of IDS specifications.

The request message types that are shown in Fig. 6 and Fig. 4 are executed by Dataspace applications (in our case, Postman tool) towards the Data Producer and Data Consumer, and these requests interact with the IDM and Metadata Broker services in their backends (via relevant front-ends, e.g. proxy URL). For example, a register connector request {PoC_EP}/api/ids/connector/update?recipient={BROKER_EP} at the data producer or consumer endpoints registers them with the provided metadata broker endpoint after it gets validated through IDM based on OAuth2 standard (using Omejdn service) workflow to enable them for cross-domain digital negotiations for data exchange. The following list presents the high-level flow of functional requests illustrated in the system architecture of Fig. 4:

- **Offer:** creates an offer/resource for the catalog at Data Producer.
- **Catalog:** creates a catalog at Data Producer.
- **Contract:** creates a contract template for the data to be used from the catalog by external parties.

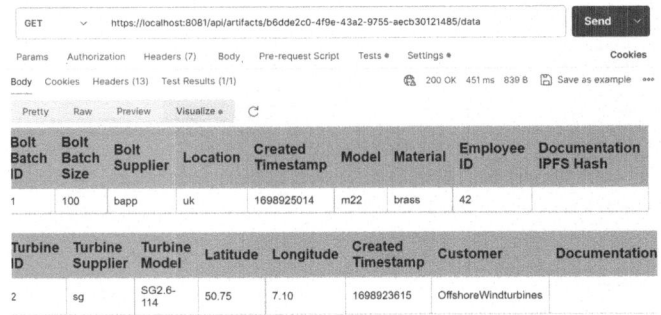

**Fig. 7.** IDS Based Cross-Domain (Consumer/Producer) Data Request

- **Rule:** creates a rule or Policies (as per IDS information and ODRL model) to be enforced for data usage during cross-domain data exchange. This rule is then added to the Contract metadata.
- **IDS Information model:** is the IDS ecosystem's default ontology or domain model to stay compliant with IDS standards and specifications to achieve interoperability and integration. Data models of all requests must comply with this model.
- **Representation:** specifies the format and language in which the artifact will be presented. For example, it could be in JSON/CSV format using English.
- **Artifact:** holds the actual data to be shared under the "*value*" field. Here, the JSON data payload needs to be encoded in string format.

During validation, entities such as offers, catalogs, contracts, rules, artifacts, and representations were generated through HTTPS POST requests at the Data Producer's end (see Fig. 6). After validation, the Data Producer and Consumer register these entities with the Metadata Broker, enabling data exchange through IDM validation. Once registered, the Data Consumer acquires metadata about the catalog from the Metadata broker and initiates direct negotiations with the Data Producer. These negotiations (trusted through IDM) involve agreeing upon and accepting the contract terms and rules stipulated by the producer. Upon successful negotiation, the consumer gains access to the artifact to retrieve the desired data. In our case, one sample of the received data, at the consumer end application, after the negotiation is shown in Fig. 7.

During deployment, we faced several challenges, which we fixed. This includes `Docker` level changes where the `version` attribute was missing in *docker-compose.yml*. Also, during the `Registration` request at MetaDatabroker, the `Omejdn` service was found to be unreachable due to using `localhost` in the service discovery URL. This was resolved by switching to a container-level URL, `broker-reverse-proxy`, representing the container-level IP, ensuring proper internal microservice discovery. Apart from that, at the environment level, we found that the `.env` file, while instantiating the docker-compose environment, using quotes, caused issues. It's recommended to omit them or use `absolute paths` for clarity and to avoid potential errors. Finally, during test validation, the `Postman` collection script encountered difficulties extracting parameter values, particularly the `Consumer Agreement Id`, from the current request to populate subsequent requests. This issue has been resolved by putting a generic code to extract parameter values, and the fix has already been pushed to the IDS main repository [41].

## 5   Conclusion

Data has become an economic asset for modern era. It is the key to the future for any industry to sustain and grow through driving data intelligence based on modern machine learning techniques and methods. Industries that operate in silos cannot exploit the true potential of their data. Therefore, the European Commission (EC) has envisioned developing cross-domain data integration (CDDI) platforms called Common Data Spaces to thrive in data-driven value chains and benefit the social, industrial, and economic factors. International Data Space (IDS) was such an initiative backed by the EC to uplift CDDI activities at regional and national levels across Europe. IDS association has developed IDS specifications, data exchange standards, and reference implementations to be used by the target industrial use case. However, the edge-driven CDDI through IDS was found to be complex due to the vast amount of scattered and hard-to-navigate information. Therefore, as a contribution, this study has developed the knowledge on the main research question of how to build IDS-based Dataspace for the target use case focusing on CDDI. As a result, we have modeled and evaluated a wind turbine industry use case leveraging the IDS information model. The platform is deployed following microservices architecture and evaluated over Docker-based virtualized infrastructure in the cloud and on-premises environment. The platform comprises identity management, data producer and consumer, metadata broker, and data source-based services that run and enable the edge-driven cross-domain Datsapce ecosystem together. As a contribution, we have advanced the relevant knowledge base on building IDS-based Dataspace at edge by applying target use case from theoretical and practical perspectives. We believe this will contribute to pace up the adoption of IDS-based standardized Dataspace implementations across industries.

## References

1. Meindl, B., Ayala, N.F., Mendonça, J., Frank, A.G.: The four smarts of industry 4.0: evolution of ten years of research and future perspectives. Technol. Forecast. Soc. Change **168**, 120784 (2021)
2. Sun, Z.: Big data 4.0= meta4 (big data)= the era of big intelligence. In: Proceedings of the 2024 7th International Conference on Software Engineering and Information Management, pp. 14–22, 2024
3. Dol, S.M., Jawandhiya, P.M.: Classification technique and its combination with clustering and association rule mining in educational data mining-a survey. Eng. Appl. Artif. Intell. **122**, 106071 (2023)

4. Allam, Z., Dhunny, Z.A.: On big data, artificial intelligence and smart cities. Cities **89**, 80–91 (2019)
5. Bachmann, N., Tripathi, S., Brunner, M., Jodlbauer, H.: The contribution of data-driven technologies in achieving the sustainable development goals. Sustainability **14**(5), 2497 (2022)
6. Qiao, L., Li, Y., Chen, D., Serikawa, S., Guizani, M., Lv, Z.: A survey on 5G/6G, AI, and robotics. Comput. Electr. Eng. **95**, 107372 (2021)
7. Pundt, H., Bishr, Y.: Domain ontologies for data sharing-an example from environmental monitoring using field GIS. Comput. Geosci. **28**(1), 95–102 (2002)
8. Sun, J., Fang, Y.: Cross-domain data sharing in distributed electronic health record systems. IEEE Trans. Parallel Distrib. Syst. **21**(6), 754–764 (2009)
9. Singh, P., Haq, A.U., Beliatis, M., et al.: Meta standard requirements for harmonizing dataspace integration at the edge. In: 2023 IEEE Conference on Standards for Communications and Networking, pp. 130–135. IEEE, 2023
10. Singh, P., Beliatis, M., Presser, M., et al.: Data-driven iot ecosystem for cross business growth: an inspiration future internet model with dataspace at the edge. In: INTERNET 2024: The Sixteenth International Conference on Evolving Internet (2024). ISBN: 978-1-68558-133-6
11. Lange, C., Langkau, J., Bader, S.R.: The ids information model: a semantic vocabulary for sovereign data exchange (2022)
12. Jarke, M.: Data sovereignty and the internet of production. In: International Conference on Advanced Information Systems Engineering, pp. 549–558. Springer, 2020
13. Scerri, S., Tuikka, T., de Vallejo, I.L., Curry, E.: Common European data spaces: challenges and opportunities. Data Spaces: Design, Deployment and Future Directions, pp. 337–357, 2022
14. Reiberg, A., Niebel, C., Kraemer, P.: What is a data space. White Paper 1, 2022
15. Nagel, L., et al.: Design principles for data spaces: position paper. Technical report, E. ON Energy Research Center, 2021
16. European Commission, A., European strategy for data. Communication COM,: 66 final. European Commission, Brussels, 2 2020
17. Lutz, S.U.: The European digital single market strategy: local indicators of spatial association 2011–2016. Telecommun. Policy **43**(5), 393–410 (2019)
18. Baloup, J., et al.: White paper on the data governance act, 2021
19. Pathak, M.: Data governance redefined: the evolution of eu data regulations from the gdpr to the dma, dsa, dga, data act and ai act. DSA, DGA, Data Act and AI Act.(February 6, 2024), 2024
20. European Commission. Implementing Regulation for High Value Datasets. https://eur-lex.europa.eu/legal-content/EN/TXT/?uri=CELEX:32023R0138, 2023. Accessed: Insert Date Here
21. Otto, B., ten Hompel, M., Wrobel, S.: Designing Data Spaces: The Ecosystem Approach to Competitive Advantage. Springer Nature (2022)
22. Singh, P., et al.: Blockchain for economy of scale in wind industry: a demo case. In: Global IoT Summit, pp. 175–186. Springer, 2022
23. Singh, P., et al.: Digital dataspace and business ecosystem growth for industrial roll-to-roll label printing manufacturing: a case study. In: SENSORCOMM 2023: The Seventeenth International Conference on Sensor Technologies and Applications. IARIA, 2023
24. Prinz, W., Rose, T., Urbach, N.: Blockchain technology and international data spaces, 2022
25. Huber, M., Wessel, S., Brost, G.S., Menz, N.: Building trust in data spaces, 2022

26. Singh, P., Beliatis, M.J., Presser, M.: Enabling edge-driven dataspace integration through convergence of distributed technologies. Internet Things 101087 (2024)
27. Torres, N., Chaves, A., Toscano, C., Pinto, P.: Prototyping the ids security components in the context of industry 4.0-a textile and clothing industry case study. In: International Conference on Ubiquitous Security, pp. 193–206. Springer, 2022
28. Volz, F., Sutschet, G., Stojanovic, L., Usländer, T.: On the role of digital twins in data spaces. Sensors **23**(17), 7601 (2023)
29. Larrinaga, F.: Bridging the gap between ids and industry 4.0–lessons learned and recommendations for the future (2024)
30. Matsunaga, I., Michikata, T., Koshizuka, N.: ITDT: international testbed for dataspace technology. In: 2023 IEEE International Conference on Big Data (BigData), pp. 4740–4747. IEEE (2023)
31. Engineering Ingegneria Informatica S.p.A. TRUE Connector. https://github.com/Engineering-Research-and-Development/true-connector. Accessed 15 May 2024
32. Pampus, J., Jahnke, B.F., Quensel, R.: Evolving data space technologies: lessons learned from an ids connector reference implementation. In: International Symposium on Leveraging Applications of Formal Methods, pp. 366–381. Springer, 2022
33. Lehtiranta, L., Junnonen, J.M., Kärnä, S., Pekuri, L.: The constructive research approach: problem solving for complex projects. Designs, methods and practices for research of project management, pp. 95–106, 2015
34. Parwinder. Wind Turbine Supply Chain Dataspace Git Repository. https://gitlab.au.dk/au656482/unwind-ids. Accessed 15 May 2024
35. International data spaces reference architecture model (ids-ram). Conceptual Framework, 2024. Version 4.2
36. Iannella, R., Villata, S.: Odrl information model 2.2. Recommendation REC-odrl-model-20180215, W3C, February 2018. W3C Recommendation
37. Hardt, D.: The OAuth 2.0 Authorization Framework. RFC 6749, October 2012. https://www.rfc-editor.org/rfc/rfc6749.txt
38. Cooper, D., Santesson, S., Farrell, S., Boeyen, S., Housley, R., Polk, W.: Internet X.509 Public Key Infrastructure Certificate and Certificate Revocation List (CRL) Profile. RFC 5280, May 2008. https://www.rfc-editor.org/rfc/rfc5280.txt
39. Chokshi, H.J., Panchal, R.: Using apache Jena Fuseki server for execution of SPARQL queries in job search ontology using semantic technology. Int. J. Innov. Res. CST **10**(2), 497–504 (2022)
40. Ibrahim, M.H., Sayagh, M., Hassan, A.E.: A study of how docker compose is used to compose multi-component systems. Empir. Softw. Eng. **26**, 1–27 (2021)
41. Singh, P.: Fixes in ids testbed. GitHub Pull Request #143, International Data Spaces Association, June 2023. https://github.com/International-Data-Spaces-Association/IDS-Testbed/pull/143

# Early Warning of Harmful Algal Blooms (HAB): A Low-Cost Integrated IoT Device with Spectrofluorometry and Automated Plankton Imaging

Gilles Orazi(✉) ⓘ, Marianne Marot ⓘ, Iheb Khelifi ⓘ, Léa Robert ⓘ,
and Franck Le Gall ⓘ

EGM, AREP Center, Traverse des Brucs, 06560 Valbonne, France
{gilles.orazi,marianne.marot,iheb.khelifi,lea.robert,
franck.le-gall}@egm.io

**Abstract.** This paper presents the development of a prototype automated device designed to monitor sea water quality and provide early warnings of potential harmful algal bloom (HAB) outbreaks. HABs pose a significant threat to aquaculture operations and marine ecosystems due to their ability to cause mass mortalities of fish and shellfish. The device integrates a low-cost custom spectrofluorometer capable of measuring absorption and fluorescence spectra of liquid samples, and an automated plankton imager adapted from an open-source design. Key aimed sensing parameters include nutrients, chlorophyll, water temperature, and phytoplankton presence. The affordability of the device is targeted by using low-cost components and integrating the spectrofluorometer and plankton imager into a single unit driven by an embedded computer. Machine learning algorithms are employed for real-time anomaly detection from the multivariate sensor data streams to provide early alerts of potential HAB events. Initial results demonstrate the device's ability to detect low concentrations of fluorescent dyes and phytoplankton, and the effectiveness of an adaptive anomaly detection approach on real aquarium data.

**Keywords:** Harmful algal bloom · Water quality monitoring · Spectrofluorometer · Automated plankton imaging · Adaptive anomaly detection

## 1 Introduction

### 1.1 The Rising Threat of Harmful Algal Blooms in Aquaculture

Over the past 50 years, the growth of industrialization and urbanization has resulted in water quality issues. Water contamination by anthropogenic substances causes changes in the physical, chemical and/or biological of the water body, resulting in undesirable or harmful effects on the biota [1]. In recent years, the aquaculture industry has faced an escalating threat from Harmful Algal Blooms (HABs), posing significant challenges to farmers and ecosystems alike. These blooms, comprised of algae species capable

© The Author(s) 2025
M. Presser et al. (Eds.): GIECS 2024, CCIS 2328, pp. 169–187, 2025.
https://doi.org/10.1007/978-3-031-78572-6_11

of producing toxins harmful to marine life and humans, have emerged as a pressing concern due to their detrimental effects on aquaculture operations and marine environments. HABs represent a particularly acute threat to the aquaculture sector due to their potential to cause mass mortalities of farmed fish and shellfish [2]. These events can lead to substantial economic losses for farmers, disrupt supply chains, and undermine food security efforts worldwide. Furthermore, the environmental repercussions of HABs extend beyond aquaculture, impacting marine ecosystems and biodiversity.

One of the key drivers exacerbating the prevalence and severity of HABs is the growing influence of anthropogenic activities, notably climate change and nutrient pollution. As global temperatures rise and weather patterns become more erratic, the conditions conducive to algal bloom formation and proliferation are becoming increasingly prevalent. Warmer sea surface temperatures, altered precipitation patterns, and changing ocean currents create fertile breeding grounds for harmful algae, facilitating their rapid growth and spread. Moreover, nutrient pollution, primarily stemming from agricultural runoff, wastewater discharge, and industrial activities, serves as a catalyst for HAB development [3]. Excessive inputs of nitrogen and phosphorus into aquatic ecosystems fuel algal growth, providing the necessary nutrients for bloom formation. In coastal regions heavily impacted by human activities, nutrient pollution acts as a significant driver of HAB outbreaks, exacerbating their frequency and intensity [1, 4].

Against this backdrop of mounting environmental pressures, the aquaculture sector faces a formidable challenge in safeguarding its operations against the deleterious effects of HABs. Traditional monitoring methods often fall short in providing timely and accurate alerts, leaving farmers vulnerable to the sudden onset of blooms and their cascading impacts. In this article, we delve into the development and application of a prototype automated device designed to monitor sea water quality and provide early warnings of potential HAB outbreaks. It aims to contribute to providing a new type of in-situ monitoring devices that are affordable, capable of operating autonomously, and able to remotely send warning messages when HAB precursors are detected.

## 1.2 IoT, Edge Computing and AI for Environmental Monitoring

In response to these challenges, innovative technologies leveraging the Internet of Things (IoT), edge computing, and artificial intelligence (AI) are emerging as potential tools to fight against HABs. By harnessing the power of advanced sensors and AI models, automated monitoring devices offer a promising solution for early detection and mitigation of HAB events [5], empowering aquaculture farmers to protect their livelihoods and marine ecosystems.

Deploying a device in the middle of the sea, even in coastal regions, is very challenging. Even if we put aside the harsh environment (salt, water, wind, sun, …), and talk only about data transmissions, the usual cellular networks may not be reliable enough, and one should consider using more robust, and less power demanding ones (like LoRaWAN for example), at the expense of a very low bandwidth. This implies that, in the case of complex sensors gathering plentiful amounts of information, processing must be done on site and that network transmissions should be kept to a minimum, essentially to alert messages. The implementation of such processing at the data collection site is called *edge computing.* The convergence of IoT and edge computing technologies is revolutionizing the data collection, analysis, and decision-making processes by enabling more efficient, real-time, and decentralized operations [6]. It offers several advantages for environmental monitoring applications, particularly in remote or resource-constrained environments such as oceans and allows to envision a device to continuously monitor the seawater to implement an alerting system. This will contribute to the goal very well stated in [7] to fulfill *"the overarching needs for HAB detection and ultimately prediction"* which is *"to have tools available that are affordable, responsive in real time, and reliable"*.

### 1.3  Monitoring Parameters for the Early Detection of HABs

HABs are complex phenomena influenced by various environmental factors which are still not well known and understood. In [7] an extensive review of the parameters of interest as well as the detection strategies and water analysis methods are discussed. Our approach considers developing a device that combines two of these strategies: monitoring the key environmental indicators on one side and directly monitoring the phytoplankton populations to see if any concerning changes are occurring. With this first prototype, among the key environmental indicators, we chose to focus on the water temperature and nutrient levels (via absorbance spectroscopy). On the other side, the plankton population is monitored using the chlorophyll level (measured using fluorescence spectroscopy) and the direct observation of the plankton at the microscopic scale. The following paragraphs elaborate on each of these indicators.

**Nutrient levels**, particularly nitrogen and phosphorus, play a crucial role in the initiation and sustenance of algal blooms. Excessive nutrient inputs from sources such as agricultural runoff, sewage discharge, and atmospheric deposition can fuel the rapid growth of algal populations, leading to the formation of HABs. By monitoring nutrient concentrations in sea water, one can identify areas of nutrient enrichment and assess the potential risk of bloom development. Early detection of elevated nutrient levels serves as a valuable indicator for implementing proactive measures to mitigate the onset of HABs.

**Water temperature** plays a dual role in HAB dynamics, influencing both the growth rate of algae and the physical conditions conducive to bloom formation. Warmer temperatures can enhance the metabolic rates of phytoplankton, promoting their growth and proliferation. Additionally, thermal stratification of water layers can create stable conditions favorable for algal accumulation near the surface, facilitating bloom development. Monitoring variations in water temperature provides valuable insights into seasonal trends and spatial patterns of HAB occurrence. Early detection of anomalous temperature regimes can serve as a warning sign for potential HAB outbreaks, prompting proactive monitoring and management strategies.

**Chlorophyll**, the pigment responsible for photosynthesis in phytoplankton, serves as a proxy for algal biomass and productivity. As HABs are composed of phytoplankton species, monitoring chlorophyll concentrations provides insights into the abundance and dynamics of algal populations in marine ecosystems. Increases in chlorophyll levels may indicate the onset of algal blooms, signaling the need for intensified monitoring and management efforts. By incorporating chlorophyll measurements into the early warning system, it would be possible to track changes in algal biomass and detect potential HAB outbreaks before they reach critical levels.

Utilizing **automated cytometry** to monitor phytoplankton presence and concentration could also provide a complementary approach to traditional monitoring methods. By gathering real-time information about phytoplankton quantities, it could be possible to detect shifts in community structure indicative of HAB formation. Certain phytoplankton species, such as dinoflagellates and diatoms, are known to contribute to HAB events. Continuous monitoring of phytoplankton communities using automated microscopy could enable early detection of these species and proactive management strategies to mitigate the impacts of HABs on aquaculture and marine ecosystems.

### 1.4 Requirements

We are thus targeting a device which is:

1. Affordable
2. Sensitive to chlorophyll concentration
3. Sensitive to nutrients concentration
4. Able to control autonomously a data acquisition process and read external probes like water temperature sensors
5. Able to process locally the acquired data to implement an alerting system
6. Able to take pictures of the phytoplankton floating in the water
7. Able to connect to a remote network and exchange data with a cloud

When one considers the price of the device (requirement 1), directly using the probes and sensors available on the market to target these parameters would be detrimental to its affordability. Most of them use the spectroscopic and fluorometric properties of the compounds to detect and assess concentrations. The cost effectiveness of our device is thus sought after by designing a generic spectrofluorometric analyzer for liquids, able to work at different wavelengths simultaneously, and replacing several probes (requirements 2 and 3). The autonomous operations (requirement 4) and processing abilities (requirement 5) are ensured by a small SoC computer. The plankton imager (requirement 6) is derived from an open-hardware and open-source plankton-imaging device. As the prototype presented here was intended to demonstrate the feasibility of such a device, the requirement 7, related to connectivity, was implemented using a Wi-Fi network which is not well suited for operations at sea but fine for laboratory investigations.

## 2  Device Design and Implementation

The embedded computer described in Fig. 1 drives the whole system. The inlet is placed inside the water, while the outlet is placed outside. This makes it possible to fill or empty the device using a single pump controlled by the computer. It is thus possible to program complex acquisition loops that include the filling or the draining of the apparatus with fresh water. Moreover, the computer interacts with the spectrofluorometer and the camera of the plankton imager to adjust the acquisition parameters and retrieve data. The computer can then communicate with the external world using its network interface. This latter could be constrained in bandwidth and may only be used for alerts and basic parameters monitoring.

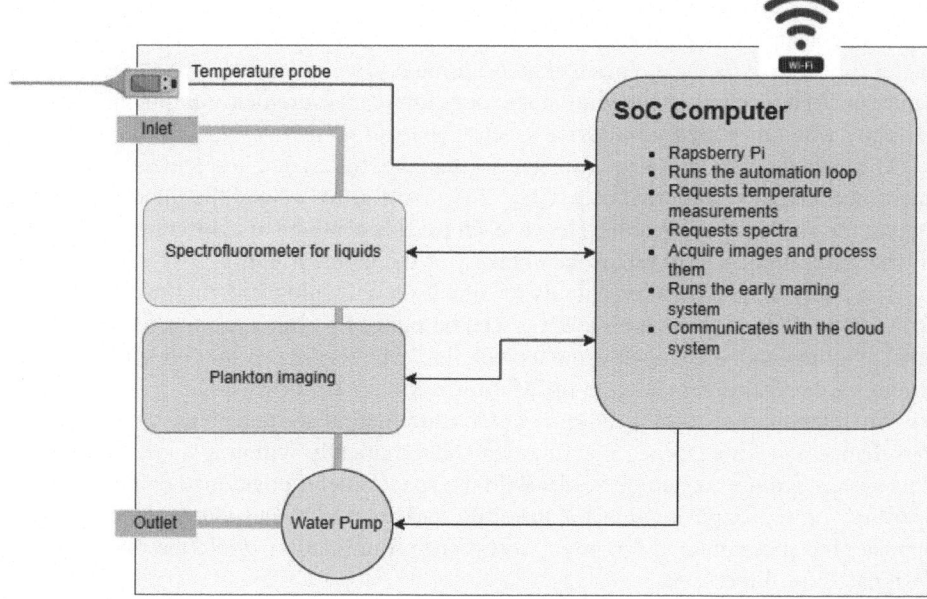

**Fig. 1.** The concept of the integrated device for fully automated warnings of Harmful Algal Blooms.

In this device, we used a DS18b20 waterproof temperature sensor; the spectrofluorometer and the plankton imaging device are described below.

## 2.1 The Spectrofluorometer

As stated in 1.4, to target a cost-effective device, we developed an affordable spectrofluorometer, which measures both absorption and fluorescence reaction spectrums of liquid compound solutions, within the near-UV, visible, and near-IR spectrum. This spectrofluorometer was developed during this study. It has the advantage of being faster than traditional manual methods, as well as enabling real-time connectivity when used as an IoT device. It integrates a miniaturized spectrometer, a sample holder designed for absorbance and fluorescence illumination, as well as a microcontroller and some interfacing electronics.

**The Mini Spectrometer**
The Micro Electro-Mechanical Systems (MEMS) processes enable the production of miniature mechanical and optical devices using techniques like those employed in the manufacturing of microelectronic chips. These attributes make them highly promising candidates as the foundation for an autonomous water analyser because of their cost effectiveness.

Our candidate for this first prototype is the Hamamatsu C12880MA mini-spectrometer due to its compact design and versatile functionality. Its spectral range covering 340 to 850 nm, even if limited to the UV range, should in principle allow seeing the absorbance of some nutrients. Its spectral resolution is about 10 nm and has a power consumption of about 0.1 Watts. Its electronic shutter ensures reliable data capture, even in dynamic environments. Additionally, weighing only 5 g, its lightweight nature facilitates seamless integration into our mobile measurement equipment, aligning perfectly with our need for in-device spectral analysis.

The principle of this spectrometer is that the optics are implemented into the micromechanical system that uses a $1 \times 288$ pixels array to read the intensities of the light at the various wavelengths. Hence, each pixel is centered at a different wavelength at which it is dedicated to report the intensity of the incident light.

The MEMS spectrometer outputs its signal on an analog line that can be read by an Analog to Digital Converter (ADC). The output of this line is connected to a single pixel at a time and is changed using a clock line. The reading of the full spectrum thus requires a driver, implemented in the Micro-Controller Unit (MCU).

We integrated it with a unique water illumination system (Fig. 2) to conduct absorbance and fluorescence measurements automatically within a single instrument. This system features a sample holder with the spectrometer positioned on one side and apertures on the opposite side for inserting various LEDs and lamps. One aperture, opposite the spectrometer, serves for absorbance illumination, while the other two are designated for fluorescence.

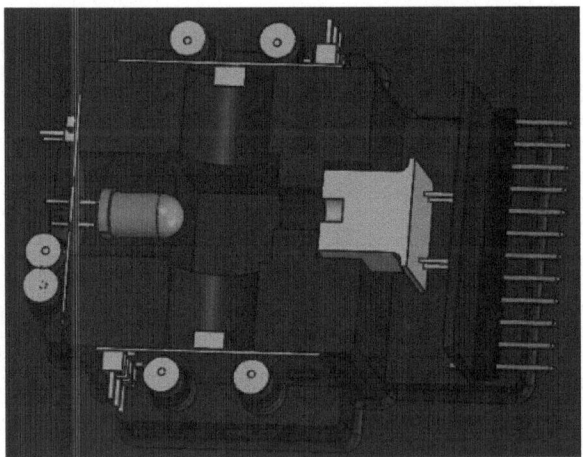

**Fig. 2.** A 3D cut view of the sample holder and illumination system. The MEMS spectrometer is the grey block on the right. On the left, in blue, one can see the lamp used for absorbance illumination and on the top and bottom two of the four LEDs used for fluorometry. The sample holder is inserted vertically in the hole in the middle of the figure. (Color figure online)

**Choice of Lamps**

Since affordability and ease of integration are key here, we have chosen to use only LEDs to illuminate the samples. The Table 1 describes the characteristics of each light source.

**Table 1.** List of the light sources (LED) used by the spectrofluorometer.

| Light source name | Excitation wavelengths |
|---|---|
| ABSORBANCE | 425–800 nm |
| FLUO-01 | 405 nm |
| FLUO-02 | 470 nm |
| FLUO-03 | 520 nm |
| FLUO-04 | 590 nm |

As one can see, the bandwidth used for the absorbance is limited to wavelengths greater than 425 nm while the spectrometer can detect as low as 340 nm. This limitation comes from the white LEDs findable on the market. In a first attempt, we tried to complement it by a UV one; however, achieving enough stability in their relative delivered light power was very challenging. We thus decided, as a first step, to continue experimenting without the near-UV spectrum, being conscious that this might negatively influence the sensitivity of the instrument to compounds such as some nutrients whose sensitivity lies at lower wavelengths than the ones detected here. However, our goal is to validate the concept of such an integrated instrument and these wavelengths complemented by the four fluorescence lights should be sufficient to detect relevant data to start HABs monitoring, especially since fluorometry at 520 nm and 590 nm are widely used to detect Blue-Green Algae compounds [8].

**Electronics**
This device is driven by a microcontroller (STM32L4) through an interfacing electronic board allowing to pilot independently each lamp and drive the spectrometer to get the data and pre-process them for background subtraction. It communicates with the computer using a simple Universal Asynchronous Receiver/Transmitter (UART) serial line.

The intensity of the LEDs is driven by a Pulse Width Modulation (PWM). The readout is done using the 16 bits ADC of the MCU.

**Embedded Software and Acquisition Procedure**
A data acquisition on this spectrofluorometer consists of taking several spectra under various illumination conditions. One single spectrum is acquired by specifying the LED to be switched on, the power level, and how long the sensor must be exposed to light (a.k.a. the integration time). Then, the MCU drives the clock and ADC lines to read the raw value of each pixel (each pixel corresponding to a specific wavelength). This latter must be corrected to dark currents and for some possibly remaining parasite lighting. This is done by acquiring a spectrum with the same exposure settings but without any light turned on and subtracting it from the raw data taken with the light on. Hence, the full acquisition procedure records a list of 5 spectra, one per lighting of the same sample. The spectrometer implements a set of high-level operations which operate autonomously (Table 2). These operations are triggered using a UART serial interface.

**Table 2.** High-level commands of the spectrofluorometer device

| COMMAND | Parameter | Description |
|---|---|---|
| GET_CONFIG | | Returns the configuration data of the full acquisition procedure, in a Json format |
| SET_CONFIG | Configuration in Json format | Sets the configuration data of the full acquisition procedure |
| START_ACQ | | Starts a new acquisition. The acquisition will be stored permanently in the device and will be associated with a unique ID |
| LIST_ACQ | | Lists all the unique IDs of the acquisitions stored in the device |
| GET_ACQ | Unique ID | Gets the acquisition data associated to a unique ID |
| DEL_ACQ | Unique ID | Deletes the acquisition data associated to a unique ID |

## Instrument Characteristics

### Noise Measurement and Control

The noise in the measurement of the intensity in each pixel limits the sensitivity of the instrument to detect small changes in the spectrum. Since we wanted to assess the performance of the MEMS spectrometer, we performed a noise analysis.

We first took 1000 spectra of light emitted by a white LED in identical conditions. The integration time is adjusted to obtain the maximum intensity just below the saturation level, providing a spectrum for which the intensity values range from 0 to maximum. For each pixel, the mean value and the standard deviation is computed over 1000 acquisitions. The standard deviation is interpreted as the noise of the pixel and the mean value as the measured value. Figure 3 shows the value of the measured noise as a function of the measured value for all the pixels. It also shows a fit by a function of the form $n = \frac{a}{m} + b$, where $n$ is the noise value, $m$ the measurement and $a$ and $b$ the two parameters. The first component of the noise is interpreted as coming from the collection of electrons by the semi-conductor, while the second part is the contribution of the electronic readout (independent of the measured value).

We chose to average the acquisition of 1000 spectra for each measurement, to lower the noise by a factor of about 32 and enhance the detection capabilities of our instrument. This number was chosen as high as possible considering that the acquisition time should remain acceptable for our experiments.

### Sensitivity to Rhodamine

Rhodamine dyes possess fluorescent properties that can be used to evaluate the spectrofluorometer's detection capacities. We tested four different concentrations of rhodamine in distilled solutions: 1000, 50, 25 and 10 µg/L. Results show that rhodamine dye emits fluorescence at wavelengths of 590–600 nm when excited at 420 nm (Fig. 4

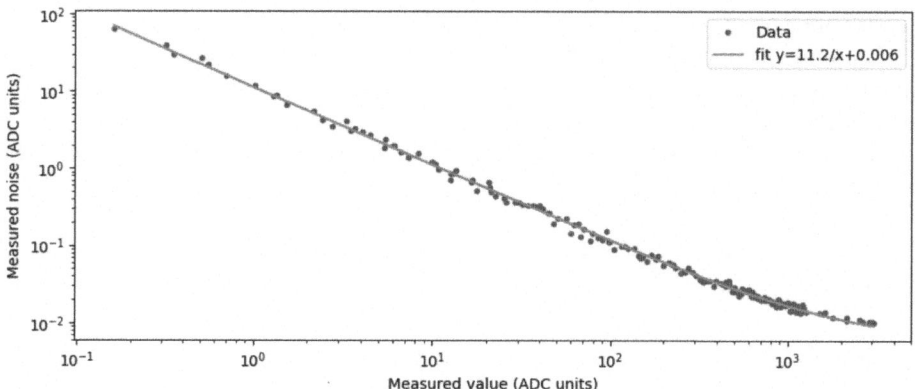

**Fig. 3.** Measured noise of the MEMS spectrometer as a function of the measured value. In orange the best fit showing the contributions of the different sources of noise (see text).

left). Furthermore, there is a positive (exponential) correlation between solution concentration and fluorescence intensity (Fig. 4 right). Hence, the spectrofluorometer functions correctly and is sensitive to concentrations of rhodamine of less than 10 μg/L.

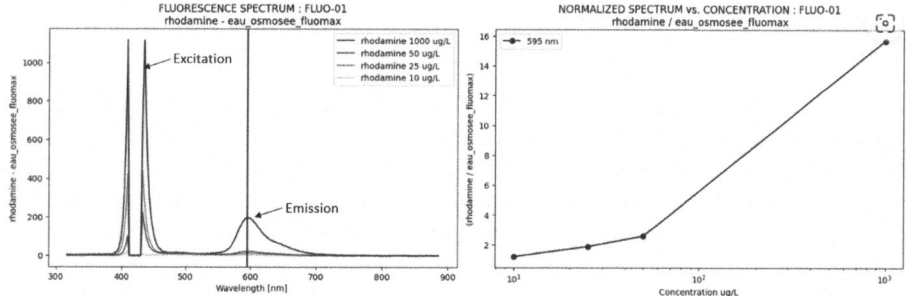

**Fig. 4.** Sensitivity test to rhodamine die for an excitation at 405 nm (left). Different concentrations of rhodamine were tested for their fluorescence emission response (right).

*Sensitivity to Phytoplankton*

To assess the sensitivity of the instrument to the phytoplankton, we used a culture comprising of five species of phytoplankton from a mixture of commercialized phytoplankton fertilizer:

– *Nannochloropsis oculate*
– *Nannochloropsis salina*
– *Dunaliella Salina*
– *Tetraselmis suecica*
– *Synechococcus SP*

It is maintained in a 2L container at room temperature, exposed to sunlight, and oxygenated with an artificial bubbler. The concentration of phytoplankton in the sample used for our experiment was not measured directly. Having no measure of the true concentration for the phytoplankton culture, we established its concentration at an arbitrary unit of 1, and then diluted it progressively with pure seawater solution and measured the relative response of the device as a function of the dilution.

We measured this dependency at four excitation wavelengths of our device. The Fig. 5 shows a typical response curve, at 420 nm (FLUO_2).

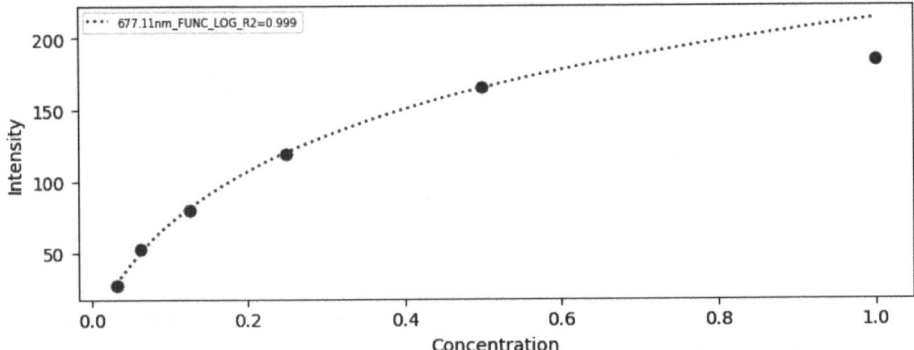

**Fig. 5.** Intensity of the fluorescence peak response at 677 nm as a function of the phytoplankton concentration. This curve was established at an excitation wavelength of 520 nm. The concentration is given in arbitrary units, starting at maximum of 1.

We noticed that fluorescence response increases logarithmically as the concentration of phytoplankton increases. However, at relative concentration of 1, data no longer fit this logarithmic model. This is explained by the high sample turbidity which absorbs photons and darkens the emitted amount of light, or due to light quenching. Indeed, [9] observed the same results, such that above a certain level of light intensity and hence amount of energy absorbed by the chloroplasts cells, phytochemical quenching and structural changes are triggered.

## 2.2 The Automated Flow Cytometer

The objective here is to provide a device that assesses the phytoplankton concentration from a liquid sample using microscope images from which simultaneous numerating of visible cells is performed. Such a device is called a flow cytometer.

We explored the possibilities of gathering such data by relying on an open-source device, called Planktoscope, designed for citizen science and specifically to provide images of plankton species within a sample of sea water [10]. It was designed to be built by an amateur scientist simply with common and easily findable parts. Since it is an open-source device, all the design and assembly instructions are available and supported by a small but very active community. It is built by assembling different functional blocks:

- Its computing module utilizes a Raspberry Pi 4 coupled with a Pi Camera v2.1
- The optical configuration includes swappable objective lenses.

- A motorized stage enables precise focus adjustment, and a peristaltic pump allows for fluid manipulation. The optical train features two inverted S-mount lenses housed in detachable modules for rapid interchangeability.
- Bright-field illumination is provided by a single ultra-bright LED, offering a large depth of field.
- The acquisition software implements a stop-flow imaging to capture stationary objects, facilitating quantitative analysis with enhanced resolution.
- The analysis software can segment the images to automatically provide cropped images of individual cells.

The Planktoscope software provides a user interface to acquire images. However, our need was to integrate it into a wider instrument. We thus needed to control it from an external process. For this, we developed a specific Python library[1].

The Fig. 6 shows some typical images of HAB plankton samples taken with our homebuilt Planktoscope. As one can see these phytoplankton cells are detectable and distinguishable by their sizes, shapes, and colors. This instrument is thus a good candidate for bringing insights about ongoing HAB developments on the monitored site.

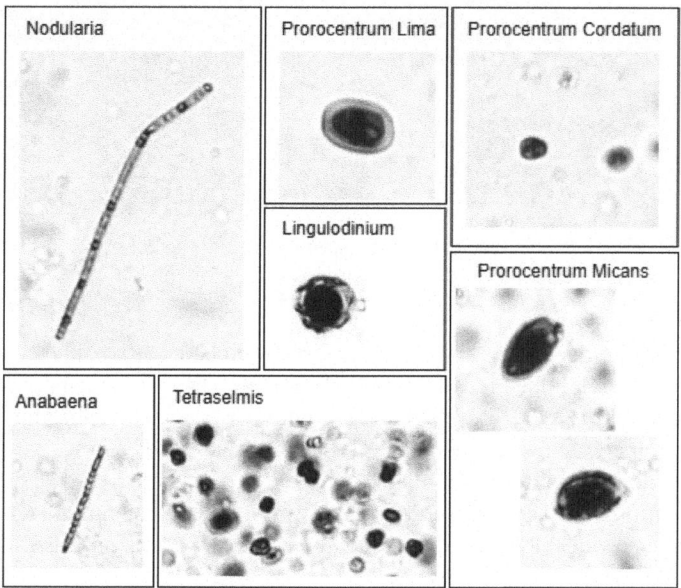

**Fig. 6.** Examples of images of HAB plankton cells taken with our Planktoscope, showing different sizes, shapes, and colors. The scale is identical for all images. (Color figure online)

---

[1] https://github.com/easy-global-market/planktomation.

## 2.3  System Integration

As stated in Sect. 1, our device prototype (Fig. 7) relies on the integration of the spectrofluorometer for liquids and the flow cytometer. Our goal is to build a single autonomous
instrument, with an automated data acquisition procedure.

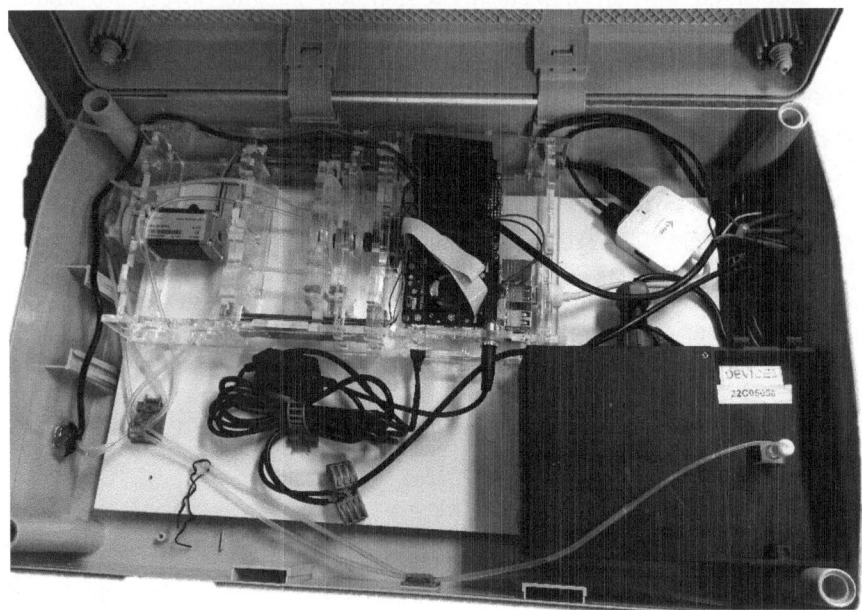

**Fig. 7.** The integrated device. One can see the spectrometer at the bottom-right (black box) and
the Planktoscope on the upper-left where the pump is at the extreme left.

The sample holder in the spectrometer was equipped with a tube with inlet and
outlets compatible with the pipes of the Planktoscope. They were then connected to the
pump so the water that exits the Planktoscope enters the spectrometer. The outlet of the
spectrometer pipe was then connected to the outlet of the whole device.

Since the SoC computer of the Planktoscope was already driving the pump and the
image acquisitions, we later added driving software to enable communication with the
spectrometer using a UART link. Consequently, it drives the entire acquisition loop such
as the following way:

1) Ensure to start with an empty tube; the pump pumps an amount of water of about the
   full capacity of the instrument in the reverse direction (outlet to inlet).
2) The same amount of water is pumped in the forward direction (inlet to outlet) to fill
   the instrument.
3) The spectrometer is then activated to take an average of 1000 spectra for each of LED
   sources. As a result, we obtain 5 spectra. Once terminated, the spectra are sent to the
   SoC computer where they are stored for later analysis.
4) A full acquisition cycle of the Planktoscope is performed, consisting of acquiring
   several images (possibly thousands) while pumping some water in between, allowing

to scan a fixed amount of water. The Planktoscope pre-processes the images to crop around the cell location. The result is stored along with the spectra for later analysis.

5) The tube is emptied by pumping in the reverse direction.

This cycle is repeated at regular time intervals for further measurements.

# 3   Adaptive Monitoring

The purpose of the adaptive monitoring is to implement in the device an algorithm that detects real-time anomalies and that autonomously adapts to the environmental conditions. To do so, we experimented with time-series data and a sub-domain of Machine Learning (ML) which considers streaming data rather than batch data. The result is an immediate alert emitted when an anomaly is detected.

Traditional ML techniques perform model training on batch data representing historical information. However, such approaches may not detect gradual changes in the sample properties since they consider an entire training dataset at once. Furthermore, they are computationally expensive, necessitating massive historical data banks often stored on expensive cloud structures. Recently, with the massive and continuous influx of IoT data, new incremental (also called online) ML methods have emerged. They can directly and quickly train ML models on continuous real-time data inflows, enabling the quick and autonomous adaptation to changes in environmental or data conditions (e.g., data drifting). Hence, they allow frequent model updating without the constraints of data storage and high computational power. Nevertheless, shortcomings of incremental learning are that it is sensitive to poor quality data and the distribution of data types (i.e., clusters of anomalies).

## 3.1   The River Tool and Selected ML Algorithm

River is a Python ML library designed for incremental state-of-the-art ML learning [11]. It is the merging of two popular stream learning packages: Creme and Scikit-Multiflow. River is particularly suited for real-time anomaly/outlier detections under changing environmental conditions, as it performs continuous model updating of its hyperparameters as new data streams in. The model can be pre-trained prior to stream-learning using historical batch data that has been cleaned from anomalies and then used and updated as new data streams in. It may also be trained from scratch in real-time. Depending on the complexity of the data, model performance is expected to be low at first, until it has learnt sufficiently to properly predict data. It is worth noting that the model will perform better when anomalous data is spread out rather than grouped, and when they are absent or rare during the early stage of the model training.

Among the set of algorithms available, HalfSpaceTrees was chosen to best perform for our purpose. HalfSpaceTrees is a tree-based algorithm for one-class adaptive anomaly detector of streaming data and is an online variant of IsolationForest algorithm. Data is expected to be standardized, and so we performed a min-max variable-wise normalization of each new data point, relative to incrementally updated statistics. Computation time is longest during the first call when the model is constructed, depending on the number and height of trees to be constructed for the model (defined by the user). Each sample of data (dictionary of key features and values) is then attributed an anomaly score between 0 and 1 (1 = anomaly). An alert is given when the anomaly score for a data point exceeds a certain threshold value, evaluated based on the data considered and context.

## 3.2 Input Data and Data Preparation.

A raw data point is composed of four fluorescence LED spectra (FLUO-01, FLUO-02, FLUO-03 and FLUO-04) and two absorption spectra, sampled at regularly wavelength intervals. Data preparation for analysis and anomaly detection is processed in the manner shown in Fig. 8.

Each raw spectrum $S_O$ (Fig. 8A) is cropped (Fig. 8B) around the reaction intervals where the fluorescence emission of the reaction occurs. The wavelengths to be cropped depend on the light source, and extract edge effects which are noisy. To emphasize these spectral responses relative to a reference spectrum $S_R$, the relative absorbance $A$ or fluorescence $F$ spectra are then computed:

$$A = -log \frac{S_O}{S_R}$$

$$F = S_O - S_R$$

where $S_R$ describes the reference spectrum chosen to reflect normal or base fluid conditions, from which the subsequent spectral responses will be compared to.

In total, 450 samples of seawater were obtained over a month of experimentation from our laboratory aquarium (representing a coral reef ecosystem). Our dataset consists of 5 variables: 3 values of fluorescence sources (FLUO-01, FLUO-02, FLUO-04) representing the maxima at relevant chlorophyll wavelength 680 nm, as well as sample temperature and phytoplankton density. For anomaly detection, we then purposely injected artificial outlier data into our database in the test sample.

### High Dimensionality Reduction by Feature Selection

The more the input features, the more challenging the modeling task becomes due to degraded computational performances. Feature selection and data condensing are two methods used to reduce our data dimensionality.

In a preliminary study of our data, we noted the similarity between certain spectral responses. To optimize computational and model training performances, we used feature selection to obtain a minimum number of features to analyze. In addition to cropping, the high dimensionality of a spectrum array itself remains problematic, comprised of hundreds of pixels, and not all necessary. Hence, in this procedure, we narrow down further the cropped spectra to their maxima values (peaks). Peak values are a strong indicator of change in fluid properties at specific wavelengths. Hence, we developed an algorithm that automatically extracts peaks from 1D-spectra to output an array of maxima intensities and wavelengths, per type of light. Often, only a single maximum is obtained per spectra, but bimodal responses can result in multi-maxima which are worth accounting for, for instance for FLUO-01.

For our anomaly detection, we considered only fluorescence reactions of FLUO-01, FLUO-02, and FLUO-04, as well as absorbance spectra 2, and we extracted the spectra maxima intensities around at the 680 nm wavelength, reflecting chlorophyll sample concentration [1]. A dictionary of key features (variables e.g. FLUO-01) and their maxima represents each data point, which is fed into the AI model to be evaluated and trained for adaptive anomaly detection.

### Normalization

Before running the model training, we first instantiated a statistics variable per feature, using Python method `runstats.Statistics`, that updates the descriptive statistics of this each feature as data streams in. We then scale each feature value using min-max normalization based on the historical data. N.B., we do not use River's normalization method, because feature values for one sample are normalized relative to one another rather than to statistics proper to each feature and updated incrementally.

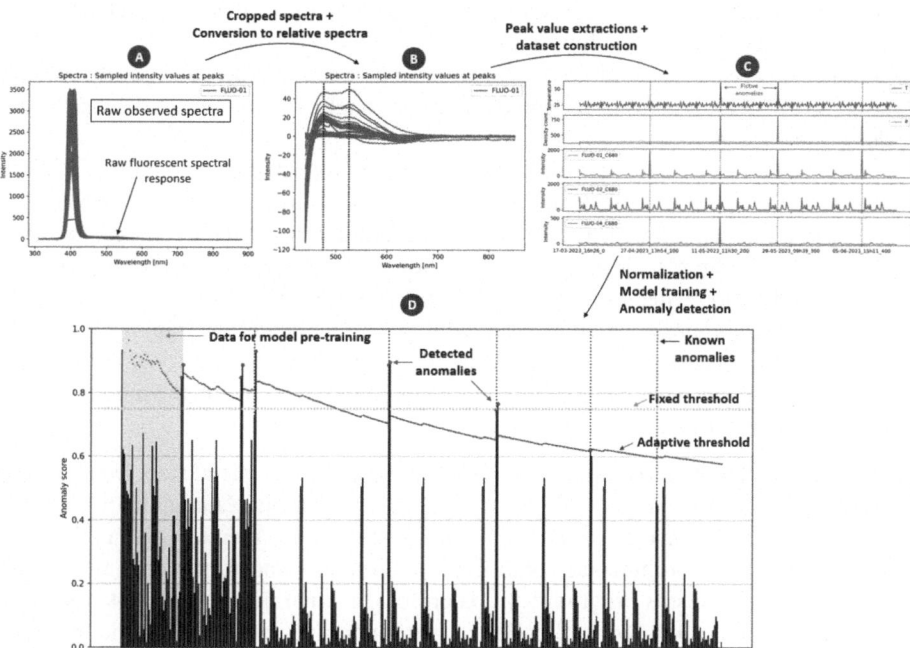

**Fig. 8.** Our stepwise procedure for data preparation of (A) raw absorbance and relative fluorescence spectra, using (B) dimensionality reduction by spectra cropping, automatic peak extraction, to create (C) datasets of time-series of spectra peaks used to fed as a streaming scenario into (D) our automatic AI anomaly detection algorithm for streaming data. The model was pre-trained on a sample of 10% of the dataset (shaded yellow region) to fine tune its parameters, and 5 fictive anomalous data points were inserted into the model using various levels of anomalousness and locations in the data. Two different methods of anomaly thresholds were used, fixed and dynamic. (Color figure online)

### 3.3 Testing Model Robustness with Real Data

We experimented using the River HalfTreeSpace algorithm to fine-tune the best model parameters (number of trees and tree height) and obtain initial scores per data sample by pre-training the model with the first 10% of normal data (45 samples) (Fig. 8D). To study the performance of model learning as data streams in and anomaly detection, we experimented with and without model pre-training.

Results show that the best parameters for the final model are 50 trees, each composed of 3 leaf levels. The model's performance to detect outliers is impacted by the number of variables representing anomalous values within a datapoint (each data point consists of 5 variables). However, we have made several observations that are worth to mention:

– No matter the importance of the outlier for a variable, the model will not detect the point as an anomaly if only 1 or 2 variables out of the 5 are anomalous; the model requires at least 60% of the point variable values to be anomalous.
– Model performance is greatly boosted when it learns rapidly from occurrences of strongly anomalous data (i.e. $\geq 60\%$ of the variables), rather than when it learns from gradually increasing anomalous data.
– Anomaly scores become more outstanding relative to the other data points, directly after model learning from a clear anomaly, strongly improving confidence in the detection.
– Anomaly scores are identical for a strong anomaly repeated throughout time. While repeating the experiment several times, we notice that the scores vary by $\pm 5\%$, indicating that a data point with a score near the detection threshold may be detected as an anomaly depending on the model run. It is therefore important to take this uncertainty into consideration when establishing detection thresholds.
– However, results hardly varied between with or without model pre-training, indicating that it is not a necessity in our case study. However, training the model with normal data allows us to determine the initial model parameters to enable the model to adapt more quickly during stream-learning.

The determination of the threshold of anomalous data is a crucial step. Using a fixed threshold value, a score threshold of 0.75 represents well our data anomalies ($\geq 60\%$ anomaly in a data point). However, a fixed threshold is ill adapted to changing data structures, due to environmental conditions for instance, and to the changing model scores as the model learns progressively. Therefore, a dynamic and adaptable threshold value improves substantially results. Hence, we found that the best dynamic threshold value, which enables correct detection of 4 out of 5 anomalies, is $\mu + 2.1\sigma$, where $\mu$ is the population mean and $\sigma$ the standard deviation. The downfall of this method is that the mean is not calculated on a sliding widow (to avoid event emphasis) and hence takes a longer time to adapt to the change in data structure, indicating that anomalous data too close to each other may not be detected properly.

### 3.4 Perspectives for Anomaly Detection

We showed that River's HalfTreeSpace algorithm works well once the adequate parameter values have been found, however, we have not been able to test our algorithm on a larger dataset, furthermore, containing naturally occurring HABs phenomena and in realistic HAB conditions, reflecting HAB precursor/onset settings with progressive mounting of chlorophyll concentrations in samples.

In addition, we wish to confront our development with other more appropriate types of streaming anomaly detection algorithms (e.g. anomaly classification algorithm ISVM), with other dynamic threshold values for anomaly detection. Also, rather than downgrading full spectrum waveform to a few values, using dimensionality reduction, we wish to experiment with other anomaly detection methods that make full use of the spectrum, using for instance incremental Generative Adaptive Networks (GANs) to create a reference synthetic spectrum of normal conditions to which data will be compared to for evaluating abnormal data.

## 4 Conclusions

This paper presents the development of an integrated IoT low-cost device that combines a custom spectrofluorometer and an automated plankton imager for the purpose of monitoring seawater quality and providing early warnings of harmful algal blooms (HABs). The spectrofluorometer is capable of measuring absorption and fluorescence spectra across visible and near-infrared wavelengths, enabling the detection of key indicators such as chlorophyll, and phytoplankton presence. The plankton imager, adapted from an open-source design, provides microscopic imaging and quantification of phytoplankton cells in liquid samples.

Initial results demonstrate the device's ability to detect and quantify phytoplankton populations, including species associated with HAB events. The implementation of an adaptive anomaly detection algorithm based on the River Python library showcases the potential for real-time monitoring and alerting of anomalous conditions that may precede or indicate the onset of HABs.

While further testing and validation are required, particularly with larger datasets and naturally occurring HAB events, this integrated device represents a promising step towards affordable and automated monitoring solutions for aquaculture operations and marine ecosystems. By harnessing the capabilities of IoT, edge computing, and machine learning, this approach empowers stakeholders with timely information and proactive strategies to mitigate the impacts of HABs, safeguarding both economic interests and environmental sustainability.

**Acknowledgments.** We would like to thank all the people from the ASTRAL project who helped us to with the design and testing of this prototype, and especially:

– Bruna Guterres and Marcelo Pias from the Federal University of Rio Grande (Brazil) for their collaboration in sharing with us HAB samples and fruitful collaborations and discussions about the replication of the Planktoscope device.

– Francisca Vermeulen from the Scottish Association for Marine Science for providing us with HAB samples

– Pauline O'Donohoe and Catherine Waters from the Marine Institute of Ireland for their help during the validation of the integrated device.

– Brett Macey from the Aquaculture Research and Development Department of Forestry, Fisheries and the Environment (Cape Town), Marissa Brink-Hull from the Department of Biological

Sciences of the University of Cape Town, and Marié Smith from the Council for Scientific and Industrial Research (CSIR) of South Africa, for their help in the validation of the imaging device.

This work was funded by the ASTRAL project which has received funding from the European Union's Horizon 2020 research and innovation program under grant agreement no. 863034.

**Disclosure of Interests.** The authors have no competing interests to declare that are relevant to the content of this article.

# References

1. Vasconcelos, H.C., Lopes, J.A., Pereira, M.J., Pinto, A.S.: Fluorescence behavior of phytoplankton blooms by time-correlated single-photon counting (TCSPC). In: Book: Fluorescence Methods for Investigation of Living Cells and Microorganisms (2020). ISBN 987-1-83968-040-0. https://doi.org/10.5772/intechopen.93292
2. Guterres, B., et al.: HAB detection within aquaculture industry: a case study in the Atlantic Area. In: 2023 IEEE 21st International Conference on Industrial Informatics (INDIN), pp. 1–6. IEEE (2023). https://doi.org/10.1109/INDIN51400.2023.10218124
3. Glibert, P.M.: Eutrophication, harmful algae and biodiversity—Challenging paradigms in a world of complex nutrient changes. Mar. Pollut. Bull. **124**(2), 591–606 (2017). https://doi.org/10.1016/j.marpolbul.2017.04.027
4. Barboza, L.G.A., et al.: Microplastics cause neurotoxicity, oxidative damage and energy-related changes and interact with the bioaccumulation of mercury in the European seabass, Dicentrarchus labrax (Linnaeus, 1758). Aquat. Toxicol. **195**, 49–57 (2018). ISSN 0166-445X. https://doi.org/10.1016/j.aquatox.2017.12.008
5. Khan, R.M., Salehi, B., Mahdianpari, M., Mohammadimanesh, F., Mountrakis, G., Quackenbush, L.J.: A meta-analysis on harmful algal bloom (HAB) detection and monitoring: a remote sensing perspective. Remote Sens. **13**, 4347 (2021). https://doi.org/10.3390/rs13214347
6. Hassan, N., Gillani, S., Ahmed, E., Yaqoob, I., Imran, M.: The role of edge computing in internet of things. IEEE Commun. Mag. **56**(11), 110–115 (2018). https://doi.org/10.1109/MCOM.2018.1700906
7. Glibert, P.M., Pitcher, G.C., Bernard, S., Li, M.: Advancements and continuing challenges of emerging technologies and tools for detecting harmful algal blooms, their antecedent conditions and toxins, and applications in predictive models. In: Glibert, P., Berdalet, E., Burford, M., Pitcher, G., Zhou, M. (eds.) Global Ecology and Oceanography of Harmful Algal Blooms. ECOLSTUD, vol. 232, pp. 339–357. Springer, Cham (2018). https://doi.org/10.1007/978-3-319-70069-4_18
8. Escoffier, N., Bernard, C., Hamlaoui, S., Groleau, A., Catherine, A.: Quantifying phytoplankton communities using spectral fluorescence: the effects of species composition and physiological state. J. Plankton Res. **37**(1), 233–247 (2015). https://doi.org/10.1093/plankt/fbu085
9. Gamayunov, E.L., Popik, A.Yu.: Dependence of fluorescence in phytoplankton on external exposures. Biofizika **60**(1), 143–151 (2015)
10. Pollina, T., et al.: PlanktoScope: affordable modular quantitative imaging platform for citizen oceanography. Front. Mar. Sci. **9**, 949428 (2022). https://doi.org/10.3389/fmars.2022.949428
11. Montiel, J., et al.: River: machine learning for streaming data in Python

# Integrating a Hybrid Lightweight Consensus Algorithm in HyperLedger Fabric

Fotis Michalopoulos(✉) , Sokratis Vavilis , Harris Niavis ,
and Konstantinos Loupos

Inlecom Innovation, Athens, Greece
{fotis.michalopoulos,sokratis.vavilis,harris.niavis,
konstantinos.loupos}@inlecomsystems.com

**Abstract.** In today's interconnected IoT ecosystems, blockchain technology acts as a facilitator for decentralization. Hyperledger Fabric (HLF) is a renounced permissioned blockchain platform designed for enterprise use. Currently, HLF supports two primary consensus algorithms: Raft and smartBFT. Despite HLF's modular nature, integrating new consensus algorithms remains a challenging and intricate process that lacks sufficient guidelines. In this work, we bridge this gap by providing a novel, comprehensive, yet practical guide to simplify the integration of consensus algorithms into HLF. Furthermore, we demonstrate our approach by integrating a new hybrid algorithm into HLF, specifically tailored for the IoT ecosystem. Compared to the current algorithms, experiments indicate promising performance and reliability.

**Keywords:** Blockchain · IoT · Hyperledger Fabric · Consensus Algorithm · Implementation Guidelines

## 1 Introduction

In the modern interconnected world, Internet of Things (IoT) devices have a pivotal role. Such devices generate vast amounts of data, from smart homes to industrial sensors, revolutionizing how we interact with our surroundings. In this setting, novel challenges arise, such as the decentralized and trusted management of devices and data. ERATOSTHENES project aims to tackle these challenges by developing a novel solution for the holistic lifecycle management of IoT devices [1,2]. To this end, the ERATOSTHENES solution utilizes blockchain technology to establish a decentralized IoT Trust and Identity Management Framework. By integrating blockchain, IoT devices can securely communicate, authenticate, and transact without the need for centralized intermediaries, ensuring both data integrity and privacy. Blockchain's immutable ledger enhances the trustworthiness of IoT networks by recording every transaction or interaction in a tamper-proof manner, fostering transparency and accountability. Together, IoT and blockchain empower diverse applications, from supply chain management to healthcare, paving the way for a more connected, efficient, and secure future [3–5].

Based on their membership status, blockchain networks can be classified as permissionless, where any node can introduce new blocks, and permissioned, where only

© The Author(s) 2025
M. Presser et al. (Eds.): GIECS 2024, CCIS 2328, pp. 188–203, 2025.
https://doi.org/10.1007/978-3-031-78572-6_12

particular nodes have permission to add blocks. Permissionless networks are ideal when openness, global participation, and high transparency are required. On the contrary, permissioned blockchains are best for environments where controlled access, privacy, performance, and compliance are critical, making them suited for enterprise environments. One of the most prominent blockchain frameworks is Hyperledger Fabric (HLF) [6], an open-source project under the Hyperledger Foundation, which enables the building of enterprise-grade permissioned blockchain networks. At its core, as with any blockchain technology, lies the consensus algorithm, which enables the decentralized coordination of nodes in agreeing to a universal state (i.e., adding a block to the chain).

Consensus algorithms are typically regarded as the cornerstone of blockchain, as they play a significant role in defining the network's security properties and efficiency. Within HLF's modular architecture, the Ordering service is in control of establishing consensus among different nodes and broadcasting newly crafted blocks to the Peer nodes. At the time of this writing, the latest HLF release, v3.0.0-beta [7], supports RAFT and smartBFT consensus algorithms. The first is a Crash Fault Tolerant (CFT) leader election algorithm, whereas the latter is a Byzantine Fault Tolerant (BFT) algorithm, recently incorporated into HLF.

Despite the modular nature of HLF and the claimed flexibility of incorporating new consensus implementations, this has been proven a strenuous endeavour. In more detail, integrating a BFT algorithm into HLF was the outcome of persistent effort, documented in several works and community discussions [8–11]. These works highlighted the challenges of consensus algorithm integration and proposed different redesigns of the ordering service. We note that even after the integration of smartBFT [10], its proposed redesign of the Ordering service is still under discussion within the HLF community [11]. Thus, the addition of consensus algorithms to HLF remains an open challenge.

Towards this direction, in this work, we provide a guide on integrating consensus algorithms into HLF, aiming to facilitate the process. In particular, we deliver a comprehensive technical description of the steps needed to add a new ordering service implementation to the current HLF architecture. Such a guide not only demystifies the integration process but also encourages prospective developers to contribute their solutions to Fabric. Furthermore, we follow the proposed approach and integrate a lightweight hybrid consensus algorithm tailored to the IoT ecosystem. The algorithm employs random lotteries and reputation-based voting to add new blocks to the chain.

The rest of the paper is structured as follows: The next chapter presents fundamental background information on consensus algorithms and HLF to assist the reader's comprehension of our work. Section 3 discusses HLF's Ordering service and the changes required for integrating a consensus algorithm, while Sect. 4 discusses the implementation details and the integration of the hybrid consensus algorithm. In Sect. 5, we perform a comparative analysis between the newly added algorithm and the already supported algorithms of HLF. Ultimately, Sect. 6 and Sect. 7 discuss the related work and conclude the paper, respectively.

## 2  Background

To assist the reader in understanding our work, this section discusses some necessary background topics.

## 2.1 Consensus in General

Consensus algorithms form the basis of blockchain technology, as they guarantee the secure storage and integrity of data on the blockchain. Specifically, consensus mechanisms provide the central mechanism that enables nodes to agree on adding a particular block to the chain in a decentralized manner.

Although several different consensus algorithms exist in the literature [12–14], they can be classified into two broad categories. Firstly, voting-based or leader-based approaches originate from the distributed systems field and rely on agreement protocols that reassemble human-like procedures. More specifically, these approaches use a voting system to elect a leader responsible for adding a block to the chain. Such algorithms are usually employed in permissioned blockchains. Both algorithms supported by HLF (i.e., Raft and smartBFT) belong to this category. On the other hand, modern consensus algorithms, typically found in public blockchains, employ randomness and competition mechanisms. For instance, in algorithms such as Proof-of-Work [15], nodes compete with each other in solving a cryptographic puzzle, and the winner adds a block to the chain. In other approaches like Proof-of-Stake [16], nodes are selected according to a probability proportional to the stake they have invested in the network. The hybrid algorithm implemented in this paper is inspired by such approaches but also utilizes a voting mechanism.

## 2.2 Fabric Overview

As already mentioned, HLF is a framework for creating permissioned blockchain networks. In HLF terminology, each different network consists of a channel, which is described as a well-defined group of nodes (i.e., consortium), that might belong to different organizations.

The network nodes serve distinct roles, being either Peers or Orderers. Peer nodes host the ledger, chaincode (i.e., smart contracts), and services such as the Fabric Gateway [17]. They have a significant role in facilitating the seamless operation of an HLF network by *endorsing* and *validating* transactions destined for blockchain inclusion. Such transactions are *submitted* to the blockchain via the Fabric Gateway. Orderer nodes are responsible for ordering endorsed transactions to blocks and propagating them to the Peer nodes. Furthermore, the cluster of Orderer nodes spanning different organizations but connected to the same channel constitutes the network's Ordering service.

Most blockchain networks adhere to an **order-execute** architecture in which the consensus protocol validates and structures the transactions forming a block. Then, the block containing the transactions is propagated to all other peer nodes. Finally, each peer node executes the transactions within the block in a sequential order. Fabric introduces an innovative **execute-order-validate** architecture. As its name implies, transactions are, firstly, executed and examined by the endorsing Peer nodes. Subsequently, the Ordering service oversees the ordering of these transactions into blocks. Finally, each transaction within the block undergoes validation by the relevant Peer nodes based on the network's endorsement policy.

The lifecycle of a transaction is initiated by a client belonging to a network's organization. The client application connects with the *Gateway* within a Peer, and then three phases occur until the transaction is appended to the blockchain.

1. In the first phase, the transaction proposal is disseminated across the network's participating organizations. During this phase, the proposed transaction needs to be executed and signed by all Peers based on the endorsement policy of the network.
2. In the second phase, the *Gateway*, based on the consensus mechanism, proceeds to either transmitting the signed transaction to an Orderer node as in the Raft implementation or broadcasting it to every member of the Ordering service as in SmartBFT. The Ordering service orders the transaction within a block and forwards the block to the *Gateway*.
3. In the third phase each Peer validates every transaction within the block. Valid transactions are committed to the ledger. The client receives a commit event from each Peer, confirming that the transaction has been successfully appended to the blockchain.

# 3   Ordering Service Documentation

In this section, we delve into the technical details related to incorporating a consensus algorithm into HLF. Note that integrating a new algorithm to HLF does not work in a plug-and-play manner (e.g., adding a new module to HLF) but requires specific changes to HLF's codebase. To this end, first, the related HLF source code should be downloaded from the official repository [18]. The repository also includes the Makefile, which is responsible for generating the necessary images and binaries used in the configuration of the Fabric network deployment (e.g., fabric samples).

To integrate a new solution, two main components need to be modified, namely the *Gateway* and the *Orderer Consensus*. Below, we present their usage and pinpoint the changes needed for the integration task.

## 3.1   Gateway

The *Gateway* component facilitates transaction submission initiated by a client application. These transactions are transmitted via the associated Peer to the Ordering service. In essence, this component handles the transactions received by the Peers of the blockchain network and, according to the nature of the consensus algorithm, handles their distribution to the Orderers.

The *Gateway* component resides within the "/internal/pkg/gateway" directory in the source code tree. The logic dictating the transaction submission process is located in the "*submit.go*" file. Within this file lies the *Submit* function, which implements the actual handling of transaction submission. Implementing the *Submit* function body is critical as it dictates the bootstrapping process of the integrated consensus algorithm. Currently, the two supported consensus algorithms provide different implementations of the *Submit* function, to better fit their nature of operation. In more detail:

- For Raft consensus, the Peer sends the transactions to a single Orderer.
- SmartBFT consensus broadcasts the transactions to all participating Orderers since implementing a BFT protocol requires withstanding any potentially malicious Orderer.

## 3.2   Orderer Consensus

The Orderer consensus component is the one actually implementing the consensus algorithm logic. It is composed of two distinct primary sub-components. The first one is the *Orderer server component* and the second is the *core of the consensus algorithm* to be integrated into HLF.

The **Orderer server's** subcomponent's main responsibility is the implementation and orchestration of three servers. Following, we present their functionality.

- The *AdminServer* is an HTTPS Handler, which, as its name suggests, is in charge of handling administrative requests from the Fabric CLI binaries. Specifically, it manages channel requests such as the 'join channel' request generated from the *peer* binary [19].
- The *OrdererServer* is a gRPC server implementing the AtomicBroadcast service and handles the communication with the Clients. In detail, this service consists of two bidirectional stream service methods, i.e. Broadcast and Deliver. This gRPC server serves as the recipient of direct gRPC requests initiated by clients, which are forwarded to their associated Peer and transmitted via the Gateway within the Peer.
- The *ClusterServer* oversees the inter-cluster communication. It implements the Cluster service which comprises the Step method. The *Step* method consumes *StepRequests* which can be either a *ConsensusRequest* or a *SubmitRequest*. The response is a *StepResponse* which is a response containing a success or failure.

The **Core Consensus** is the part where the business logic of a consensus algorithm is injected. Current implementations of Raft and smartBFT are located in /Orderer/consensus directory. Below the most important structs are presented.

- The *Consenter* interface needs to implement the body of the function's signature *HandleChain*. The consenter struct can also act as a wrapper for the Chain interface responsible for some configuration functionality. In the Raft paradigm, the Consenter integrates the dialer client towards the other Orderers that consist of the ordering service, along with the gRPC server handling the cluster's gRPC services. SmartBFT's implementation follows this design as well. The consenter struct contains an instance of the *ClusterServer* described previously. This service's source code is implemented in the common Orderer consensus code. The *Step* function is responsible for forwarding the request received to the relevant Chain's service implementation. The *Consenter* contains an instance of a dialer, also provided from the common Orderer code, and enables establishing connections with all other members of the ordering service. Last, a Consenter may function as a chain selector as it can integrate multiple chain instances.
- The *HandleChain's* inputs consist of a Consenter Support interface and metadata. The names of these inputs imply their functionality. It is worth mentioning that the Consenter Support interface is implemented by the ordering service common code defined in the official repository of the HLF. This code is part of the suggestion of the smartBFT paper for a more concrete implementation of the ordering service. This part of the code needs to be refactored and changed position to support an ordering service that will act as an agnostic framework to the HLF configuration.

The metadata is part of the protofiles used. The protofiles used by the ordering service are documented in a separate GitHub repository under the HLF [20]. Last, the *HadleChain's* output is the *Chain* interface.

– Lastly, the *Chain* interface is responsible for implementing the core functionality of the consensus algorithm supported by the ordering service. It oversees the functionality of the services described in the previous section. Upon receiving a request, these services forward it to the Chain's functions. The chain function that needs implementation will be explained thoroughly in Sect. 4, where our Implementation details are unveiled. The chain struct contains the node that is responsible for inter-functionalities and inter-communication between the Orderers that comprise the ordering service. The communication between the Orderers is achieved via the RPC interface provided by the cluster package. The node contains an RPC instance to facilitate communication with the other Orderer nodes.

## 4 Integrating a Lightweight Reputation-Based Consensus Algorithm

To address the needs and the limited resources of devices in the IoT ecosystem, we implement and integrate a hybrid consensus algorithm with HLF. The algorithm, semantically similar to Proof-of-Luck [21], employs random lotteries for block proposals. In addition, a consortium of trusted nodes votes for the best proposal to be added to the chain, similarly to Algorand [22], although using reputation instead of stake. This approach enhances the security of the process and assists in reaching finality. Next, we discuss the details related to the algorithm's implementation.

### 4.1 Preliminaries

Before discussing the details of the consensus algorithm, we need to present the foundations on which the solution is based.

**VRF:** The algorithm employs Verifiable Random Functions to enable the random lotteries for proposing blocks. VRFs are cryptographic functions introduced by Micali et al. [23] and have been employed in other consensus algorithms, such as the Algorand Pure PoS [22]. The inputs of a VRF consist of a public/private key pair and a seed. Upon processing these inputs, the VRF generates an output in the form of a hash. A VRF's outcome correctness is verifiable by anyone using the corresponding public key. Thus, only the private key holder can compute a VRF hash, but everyone can verify it. The output of the VRF is uniformly distributed as it is the result of the private key and the seed in a unique manner, amplifying randomness [24].

**Reputation:** Another essential part of the integrated algorithm is reputation. In particular, in this approach, each participating node has a reputation value, fostering a sense of trust among applicants. Typically, to assess the reputation of nodes, reputation systems [25–27] are used. Such systems employ diverse methods to estimate the trustworthiness level of entities based on their previous behaviour. In this work, we assume that a reputation system already exists; thus, each node's reputation is accessible from the implemented algorithm. Hence, trusted Orderer nodes (i.e., having high reputation) are named *reputable* and participate in the voting phase. On the other hand, non-trusted nodes are called *non-reputable*.

## 4.2  Algorithm Implementation Details

The implemented algorithm diverges from existing consensus approaches in HLF by not relying on a leader-based model. Specifically, block proposals may originate from any network node via random lotteries (i.e., no leader selection phase), while a voting phase only occurs among reputable nodes to ensure finality. Its hybrid lightweight design lacks resource or energy-intensive processes, making it applicable to the limited resources of devices in the IoT context. The consensus algorithm consists of three phases, namely *Proposal, Voting and Commit.*

**Proposal Phase:** This phase initiates when an Orderer has received and ordered a batch of transactions, forming a block within the context of HLF. We note that the transactions inserted in the proposed blocks are removed from the Orderer's transaction pool. Each Orderer proposes the self-ordered block to be appended to the shared ledger. Next, each Orderer calculates a VRF associated with the block using its private key and the seed.

---

**Algorithm 1:** Proposal Phase

---

calculate self hash;
bestVRF = self hash;    // The bestVRF initialized with self hash and will hold the best VRF received from the Orderers
broadcast self hash to other Orderers;
**while** *Proposal Phase timer NOT expired* **do**
  receive a vrf request;
  **if** *vrfReceived verified against Orderer's public key* **then**
    **if** *vrfReceived > bestVRF* **then**
    └ bestVRF = vrfReceived
    blockProposals ++;
    **if** *blockProposals == majorityOfOrderers* **then**
    └ break;

---

The seed used in the VRF can be extracted from the last block's header of the ledger. This enables any Orderer node to verify the generated VRF hash using the block's seed and the respective orderer's public key. The seed of each block is calculated as the VRF hash chosen for the specified block to enter the ledger. For the sake of consistency, the first block to the ledger will have VRF calculated from a genesis seed, which will be randomly calculated for each system serving as the bootstrap seed.

After block preparation, each Orderer broadcasts its proposal to the other Orderers and listens for others' block proposals. Note that, in parallel, each Orderer also listens for transactions coming from the Peers and stores them in the transaction pool. The process is summarized in Algorithm 1.

**Voting Phase:** This phase determines which block will be added to the shared ledger. The key participation criterion in this phase is each Orderer's reputation. In particular, only *reputable* nodes are permitted to participate in the voting process. The algorithm describing the aforementioned phase is presented below. At the end of the *Voting Phase* every Orderer should have a winning block.

---

**Algorithm 2:** Voting Phase

---

**if** *reputable node* **then**
    send the bestVRF to every other Orderer;
    quorum votes = quorum votes - 1; `// Not to count itself in the number of votes required`

**while** *Voting Phase timer NOT expired AND NOT quorum votes received* **do**
    receive vote;
    votes ++;                      `// Counter for number of votes received`
    **if** *votes == quorum votes* **then**
        quorum votes received = true;    `// Received the required number of votes`

Check if a proposed block has the majority of votes;    `// Find the winning block`

---

**Commit Phase:** During this phase each Orderer commits the winning block to the shared ledger. Upon successful validation of the proposed block, indicating consensus among all Orderers, each Orderer updates its state in preparation for the next round of the consensus algorithm.

In particular, each Orderer node has to update its transaction pool and prepare for the next round (i.e., block proposal). We distinguish two cases: a) the accepted block originating from the node itself (i.e., self-proposed block) and b) the accepted block originating from another Orderer. In the first case, the Orderer takes no further action. If any transaction remains in the transaction pool, the Orderer proceeds to the next round and proposes the next block. Otherwise, if the accepted block originated from another Orderer, the node must update its transaction pool by removing the processed transactions. Firstly, the Orderer cleans its self-proposed block by removing from the pool any transaction in common with the accepted block. The remaining transactions of the self-proposed block are returned to the pool. Next, the remaining transactions of the accepted block are removed from the transaction pool.

We note that the accepted block might contain transactions that are not known to the Orderer (i.e., not in the transaction pool), leading to potential duplicate transactions in future blocks. To prevent this, if a transaction of the accepted block is not yet known by an Orderer, it is temporarily stored and used for crosschecks at the latter stages. The Commit Phase logic is shown in Algorithm 3.

---

**Algorithm 3:** Commit Phase

---

```
transactionPool;                    // The received - non processed- txns
proposedTxns;              // The proposedTxns from the winning block
selfBlockTxns;                      // The txns of self proposed block
// removing duplicate txns
for txn in selfBlockTxns do
    if txn in proposedTxns then
        remove txn from the transactionPool;
        remove txn from the proposedTxns;
    else
        append txn to the transactionPool;

for txn in proposedTxns do
    if txn in transactionPool then
        remove txn from the transactionPool;
    else
        store txn in txnsMinted ;

txnsExtracted = 0;          // Counter of txns extracted from the txn pool
txnsMinted;     // The minted txns but not yet received from the Orderer
// check if there are any pending txns
for txn in transactionPool do
    if txn in txnsMinted then
        remove txn from txnsMinted;
        remove txn from transactionPool;
    else
        txnsExtracted ++;
        send transaction to the function responsible for cutting the next block;
        if txnsExtracted == batch size of block then
            break;
```

---

### 4.3 Additional Integration Details

To integrate the implemented algorithm into HLF, further modifications to the source code are necessary.

- First, the ***ConsensusRequest*** message declared in the protofiles of the Step service has been altered. We added two more fields to support the algorithm logic The first is the VRF hash that needs to be sent from one Orderer to another. Following, the NodeState field has also been added to associate the *ConsensusRequest* either to the *Proposal Phase* or the *Voting Phase* of the algorithm.
- Next, the ***Block Header*** message declared in the *common.proto* file has been altered with the addition of an extra field, namely *seed*, to denote the hash used for the specific block.

- Regarding the **public/private** keypair required for the VRF implementation, we leveraged the existing key pairs used by the *Membership Service Provider* (MSP) within HLF.
- Finally, the **gateway** package has also been modified. Specifically, the *Submit* function has been modified to support the broadcasting of the transaction to all the Orderers of the Ordering service.

## 5 Benchmarks

Upon integrating the new consensus algorithm into HLF, we moved forward with assessing its performance. The main objective of our experiments was to compare our algorithm's performance and behavior with the existing consensus algorithms of HLF, namely Raft and smartBFT. Toward this direction, we employed a variety of settings to achieve this objective, which are discussed as follows.

### 5.1 Setup

Our experiments were performed in a Virtual Machine setting, with the following specifications:

- a 24-core AMD EPYC 7452 x86-64 CPU
- 64 GB RAM
- SSD storage

For benchmarking, we aimed to use a standardised and reproducible approach. To this end, we employed the Caliper tool provided by the Hyperledger Foundation [28]. Caliper is a blockchain benchmarking solution designed to allow users to measure the performance of a specific blockchain implementation. To evaluate the performance, we configured Caliper to create a pool of 15000 transactions and use a client application to send them at a rate of 1500 transactions per second. These transactions were broadcasted from 24 concurrent workers to each Orderer node.

We conducted the benchmarks following the methodology outlined in the smartBFT work [10] and created cases with different numbers of Orderers. Moreover, the cases consist of clusters and transaction batches of various sizes. In particular, the clusters used for smartBFT consisted of 4, 7, and 10 Orderers, while 5, 7, and 11 Orderers were used for Raft and hybrid consensus. Batch sizes of 250, 500, and 1000 transactions per block were used for each cluster size.

### 5.2 Results and Discussion

For each of the benchmark cases mentioned previously, a set of 6 repetitions was executed. The benchmark results for the hybrid consensus algorithm, Raft, and smartBFT are illustrated in Figs. 1, 2, and 3, respectively. Each column in the figures is the average value calculated from the aforesaid repetitions.

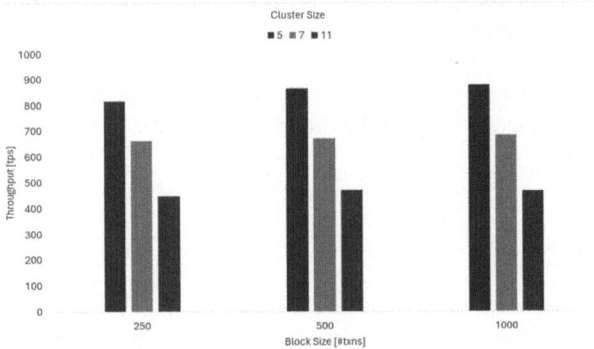

**Fig. 1.** Hybrid consensus Lan

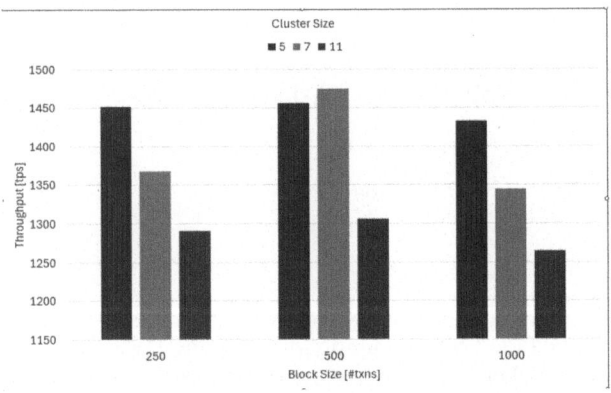

**Fig. 2.** Raft Lan

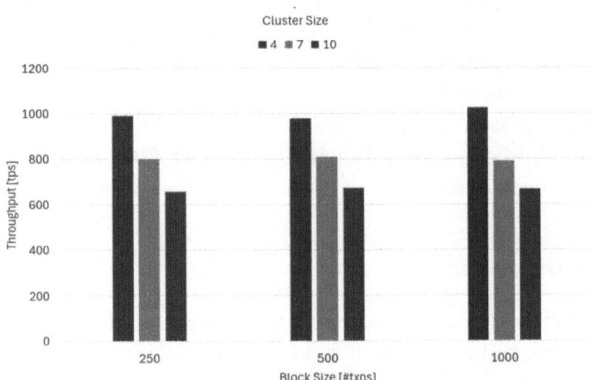

**Fig. 3.** Smart BFT Lan

Overall, Raft demonstrates superior performance (i.e., throughput) compared to smartBFT and the hybrid consensus approach. This is justified by the comparatively less network communication overhead introduced by Raft. In particular, being a leader-based algorithm, once a leader is elected, they decide on the blocks to be added and communicate such a block to their peers. Consequently, the transaction lifecycle is significantly shorter, with only one phase invoked and minimal communication between nodes. Similarly, smartBFT follows a leader-based approach but introduces a communication overhead for validating messages (Pre-prepare, Prepare, and Commit phases). The hybrid consensus algorithm diverges from this paradigm, and albeit fairer, it requires additional message exchanges between the nodes for proposing blocks and voting.

SmartBFT appears to have a slight advantage over our implementation in every case. However, it is essential to mention some observations based on the evaluation results:

- In terms of reliability, smartBFT experienced transaction failures as the number of Orderers increased. These failures ranged from a minimum of 0 to 17 in the 7 Orderers case. For 11 Orderers, the failures increased to a range between 0 and 226 (i.e., 1.5% of transactions). We have also observed some edge cases where over 2000 failures (13% of the transactions) occurred; however, we decided not to consider such instances in our comparison. On the other hand, both Raft and the hybrid consensus reported no transaction failures, signalling high reliability.
- Another less significant observation is that SmartBFT tended to generate a significantly higher number of duplicate transactions compared to Raft and the hybrid consensus. In particular, we noticed that in some cases with 11 Orderers, more than 200 duplicate transactions have been recorded.
- It is worth mentioning that despite configuring the batch size to a particular value (e.g., 1000 transactions per block), a substantial proportion of committed blocks in smartBFT contained significantly fewer transactions than the specified limit. Thus, contrary to the other algorithms, smartBFT treated the batch size as a theoretical maximum. Although such a practice may have a positive impact on the throughput of the algorithm, it makes the fair comparison of the approaches problematic, as, for instance, the hybrid consensus lacks such optimizations in its implementation.

An additional observation is related to the impact of the batch size on the throughput of the algorithms. In more detail, increasing the number of transactions per block from 250 to 500 has a positive effect on almost all the tested cases. Moreover, contrary to the other algorithms, the hybrid algorithm's approach increased its performance for a batch size of 1000 transactions, whereas smartBFT experienced a minor decrease, and Raft showed a more significant decline. Last but not least, we also noticed that the cluster size (i.e., the number of Orderers) negatively impacts the throughput of algorithms. Such a decrease is mainly a result of additional communication overhead. A notable exception is the Raft case for 500 transactions batch size and 7 Orderers.

In summary, we have demonstrated the successful integration of a new consensus algorithm to HLF. While our implementation showed promising performance results, it is apparent that further optimization is needed. Nonetheless, the tested implementation is still in the proof-of-concept stage rather than production-ready. Our next steps will focus on minimizing communication costs (e.g., reducing message size). Besides that, we aim to improve the algorithm's implementation by incorporating mechanisms to handle node failures and post-initialization changes. In conclusion, contrary to other approaches, our implemented solution scales reliably without any reported transaction failures, showcasing its real-world application potential.

## 6  Related Work

In its early versions, HLF utilized Apache KAFKA [29] as its Ordering service (i.e., consensus mechanism). This algorithm followed a "leader-follower" model and has Crash-Fault Tolerant (CFT) properties. However, it was deprecated in the later versions, mainly due to significant administrative overhead, which discourages HLF's adoption by new users. For the sake of completeness, we mention that a SOLO ordering service, consisting of only one node, was also supported for testing purposes.

Followingly, the RAFT consensus algorithm was chosen as the successor of the KAFKA Ordering service, preserving the same CFT properties while introducing an easier-to-grasp and manage solution [30]. Raft implements a leader-based model, where the leader is elected among ordering nodes participating in a channel. The elected leader node is in charge of managing state changes and disseminating them to the follower nodes. In the event of a leader's crash, a new election is initiated to appoint a new leader. For any HLF network with a channel to operate correctly, the presence of a leader node is crucial.

Third-party implementation and integration of consensus algorithms in HLF, however, have proven more challenging than initially anticipated. In particular, a few efforts took place to integrate a consensus algorithm supporting Byzantine Fault Tolerance (BFT). On this track, one of the first works was presented by Sousa et al. in 2018 [8], which described a BFT Ordering service for HLF. Nonetheless, this work was not integrated into HLF samples, as it was not adopted by the HLF community [9]. The main reason for this was the architectural misalignment of the proposed solution with HLF. More specifically, the solution proposed an external monolithic BFT cluster. An elaborate discussion on this matter can be found in [10].

Moving forward, the latest BFT implementation, which was accepted and integrated into the official HLF sample, is smartBFT [10]. Next to the integration into HLF, the work in [10] highlighted the challenges associated with implementing a consensus algorithm for HLF and proposed a novel approach for designing future ordering services. In this design, each Ordering service will work in a "plug-and-play" manner, allowing users to incorporate consensus algorithms during HLF's network deployment configuration. In particular, the authors envisioned a detachable Ordering service represented by an interface. Each new algorithm would have to implement the proposed interface and then be easily used as a plugin. Albeit very promising, such a change in HLF's Ordering service design requires significant effort and approval by the community. At the time of this writing, the proposal is still under discussion in fabric RFCs [11].

Contrary to the redesign solution presented in [10], this work follows a more pragmatic approach. In particular, we discuss HLF internals and present guidelines on implementing and integrating consensus algorithms into HLF as it is. Moreover, we demonstrate our approach by integrating a novel hybrid consensus algorithm. Our ambition is to motivate and enable developers to incorporate new algorithms into HLF, effectively enriching the HLF ecosystem.

## 7 Conclusions

In this work, we presented a systematic approach to address the challenge of integrating new consensus algorithms into HLF. To this end, we provide a guide highlighting the technical details that shall be altered to incorporate a new algorithm into HLF. Furthermore, we demonstrated the feasibility of the proposed approach by integrating a novel hybrid consensus algorithm into HLF, and we compared its effectiveness to the two existing algorithms of HLF. We believe that the insights provided in this paper will help developers extend HLF's capabilities by integrating additional consensus algorithms into the framework. Besides improving the implementation, as future work, we aim to apply it in the resource-constraint ecosystem of the ERATOSTHENES project.

**Acknowledgments.** This work has received funding from the European Union's Horizon 2020 research and innovation programme under grant agreement no 101020416. The authors acknowledge the research outcomes of this publication belonging to the ERATOSTHENES (101020416) project consortium.

## References

1. Loupos, K., et al.: Lifecycle management of IoT devices via digital identities management and distributed ledgers solutions. In: Proceedings of the Embedded World Conference 2024 (2024)
2. Loupos, K., et al.: An inclusive lifecycle approach for IoT devices trust and identity management. In: Proceedings of the 18th International Conference on Availability, Reliability and Security (2023)
3. Saberi, S., Kouhizadeh, M., Sarkis, J., Shen, L.: Blockchain technology and its relationships to sustainable supply chain management. Int. J. Prod. Res. **57**(7), 2117–2135 (2019)
4. Haleem, A., Javaid, M., Singh, R.P., Suman, R., Rab, S.: Blockchain technology applications in healthcare: an overview. Int. J. Intell. Netw. **2**, 130–139 (2021)
5. Gad, A.G., Mosa, D.T., Abualigah, L., Abohany, A.A.: Emerging trends in blockchain technology and applications: a review and outlook. J. King Saud Univ.-Comput. Inf. Sci. **34**(9), 6719–6742 (2022)
6. Hyperledger Foundation: HyperLedger Fabric (HLF) Documentation (2024). https://hyperledger-fabric.readthedocs.io/en/latest/whatis.html. Accessed 8 Aug 2024

7. Hyperledger Foundation: HyperLedger Fabric v3.0.0-beta Code repository (2024). https://github.com/hyperledger/fabric/releases/tag/v3.0.0-beta. Accessed 8 Aug 2024

8. Sousa, J., Bessani, A., Vukolic, M.: A byzantine fault-tolerant ordering service for the hyperledger fabric blockchain platform. In: Proceedings of the 48th Annual IEEE/IFIP International Conference on Dependable Systems and Networks (DSN) 2018, pp. 51–58. IEEE (2018)

9. HLF Mailing List: HLF community discussion regarding byzantine fault tolerance (2024). https://lists.hyperledger.org/g/fabric/topic/17549966#3135. Accessed 8 Aug 2024

10. Barger, A., Manevich, Y., Meir, H., Tock, Y.: A byzantine fault-tolerant consensus library for hyperledger fabric. In: Proceedings of the IEEE International Conference on Blockchain and Cryptocurrency (ICBC) 2021, pp. 1–9. IEEE (2021)

11. Hyperledger Foundation: Hyperledger Fabric ordering service consensus agnostic framework (2024). https://hyperledger.github.io/fabric-rfcs/text/Orderer-v3.html. Accessed 8 Aug 2024

12. Azbeg, K., Ouchetto, O., Jai Andaloussi, S., Fetjah, L.: An overview of blockchain consensus algorithms: comparison, challenges and future directions. In: Advances on Smart and Soft Computing: Proceedings of ICACIn 2020, pp. 357–369 (2021)

13. Stefanescu, D., Montalvillo, L., Galán-García, P., Unzilla, J., Urbieta, A.: A systematic literature review of lightweight blockchain for IoT. IEEE Access **10**, 123138–123159 (2022)

14. Lashkari, B., Musilek, P.: A comprehensive review of blockchain consensus mechanisms. IEEE Access **9**, 43620–43652 (2021)

15. Nakamoto, S.: Bitcoin: a peer-to-peer electronic cash system (2008). https://bitcoin.org/bitcoin.pdf

16. Nguyen, C.T., Hoang, D.T., Nguyen, D.N., Niyato, D., Nguyen, H.T., Dutkiewicz, E.: Proof-of-stake consensus mechanisms for future blockchain networks: fundamentals, applications and opportunities. IEEE Access **7**, 85727–85745 (2019)

17. Hyperledger Foundation: HyperLedger Fabric (HLF) v2.5 Documentation (2024). https://hyperledger-fabric.readthedocs.io/en/release-2.5/network/network.html#what-is-a-blockchain-network. Accessed 8 Aug 2024

18. Hyperledger Foundation: HyperLedger Fabric (HLF) source code (2024). https://github.com/hyperledger/fabric. Accessed 8 Aug 2024

19. Hyperledger Foundation: Create a Channel in Hyperledger Fabric (2024). https://hyperledger-fabric.readthedocs.io/fa/release-2.5/create_channel/create_channel_participation.html. Accessed 8 Aug 2024

20. Hyperledger Foundation: Hyperledger Fabric gRPC Service Definitions (2024). https://github.com/hyperledger/fabric-protos. Accessed 8 Aug 2024

21. Milutinovic, M., He, W., Wu, H., Kanwal, M.: Proof of luck: an efficient blockchain consensus protocol. In: Proceedings of the 1st Workshop on System Software for Trusted Execution, pp. 1–6 (2016)

22. Chen, J., Micali, S.: Algorand: a secure and efficient distributed ledger. Theor. Comput. Sci. **777**, 155–183 (2019)

23. Micali, S., Rabin, M., Vadhan, S.: Verifiable random functions. In: Proceedings of the 40th Annual Symposium on the Foundations of Computer Science (FOCS 1999), pp. 120–130. IEEE (1999)

24. Chainlink: Verifiable Random Functions (VRF) in simple terms (2023). https://chain.link/education-hub/verifiable-random-function-vrf. Accessed 8 Aug 2024

25. Vavilis, S., Petković, M., Zannone, N.: A reference model for reputation systems. Decis. Support Syst. **61**, 147–154 (2014)

26. Fortino, G., Fotia, L., Messina, F., Rosaci, D., Sarné, G.M.: Trust and reputation in the internet of things: state-of-the-art and research challenges. IEEE Access **8**, 60117–60125 (2020)

27. Aaqib, M., Ali, A., Chen, L., Nibouche, O.: IoT trust and reputation: a survey and taxonomy. J. Cloud Comput. **12**(1), 42 (2023)
28. Hyperledger Foundation: Hyperledger Caliper benchmarking tool (2023). https://www.hyperledger.org/projects/caliper. Accessed 8 Aug 2024
29. Hyperledger Foundation: HLF ordering service implementations (2024). https://hyperledger-fabric.readthedocs.io/en/release-2.5/Orderer/ordering_service.html. Accessed 8 Aug 2024
30. Ongaro, D., Ousterhout, J.: In search of an understandable consensus algorithm. In: Proceeding of the 2014 USENIX Annual Technical Conference (USENIX 2014), pp. 305–319 (2014)

# Author Index